Interfacial Fluid Mechanics

Vladimir S. Ajaev

Interfacial Fluid Mechanics

A Mathematical Modeling Approach

 Springer

Dr. Vladimir S. Ajaev
Department of Mathematics
Southern Methodist University
Dallas, Texas
USA
ajaev@mail.smu.edu

ISBN 978-1-4899-9896-5   ISBN 978-1-4614-1341-7 (eBook)
DOI 10.1007/978-1-4614-1341-7
Springer New York Dordrecht Heidelberg London

Printed on acid-free paper

Springer is part of Springer Science+Business Media (www.springer.com)

Preface

This book is based on advanced fluid mechanics course taught by the author at Southern Methodist University, primarily for graduate students in applied mathematics and mechanical engineering. The main focus of the course has been on fluid mechanics of microscale systems with liquid–gas, liquid–liquid, and liquid–solid interfaces. An introductory graduate-level fluid mechanics course was a pre-requisite, so students were expected to be familiar with the fundamentals of fluid mechanics of single-phase systems. The focus of the present book is on mathematical modeling of interfacial flows rather than on various sophisticated experimental techniques devised to investigate these flows on the microscale. The author's objective is to provide detailed descriptions of mathematical techniques and working numerical codes and thus make it easier for an applied math or engineering graduate student to start their research work in the field of interfacial fluid mechanics.

The interest in interfacial phenomena has been motivated by a number of practical applications of fluid mechanics in microscale systems. Remarkable examples of such systems include portable "lab-on-a-chip" devices for conducting blood tests using a small droplet of blood or identification of a virus by specific genetic sequences in a small sample containing a roughly purified DNA. Excellent discussions of these and many other applications can be found in several recently published books mentioned in the end of Chap. 1. By choosing a certain set of topics for this introductory book on the subject, the author does not attempt to state that these represent the most important problems in microscale fluid mechanics. Instead, the choice is often guided by how suitable a certain physical problem is for illustrating a mathematical method with a wide range of applicability. In some cases, the choices were also biased by the author's own research interests. Sections which contain advanced material are marked by asterisks (*). Several topics from interfacial fluid mechanics, most notably theories of interfacial instabilities, are covered in standard graduate textbooks on fluid mechanics and therefore received relatively little attention in the present book.

Many parts of the book benefited from numerous illuminating discussions with my scientific mentors, Professors S.H. Davis and G.M. Homsy, and with a number of wonderful collaborators from all over the globe, especially D. Brutin, T. Gambaryan-Roisman, E.Ya. Gatapova, O.A. Kabov, P. Stephan, L. Tadrist, and O.I. Vinogradova. I would also like to acknowledge many stimulating discussions of numerical methods of applied mathematics with my colleagues at SMU, especially I. Gladwell and J. Tausch. I am deeply grateful to several colleagues who took time to read various parts of the book. They are Steffen Hardt, Bud Homsy, Rouslan Krechetnikov, David Willis, and several anonymous referees. Many current and former SMU graduate students, especially Jill Klentzman and Christiaan Ketelaar, provided useful feedback on the manuscript.

Dallas, TX   Vladimir S. Ajaev

Contents

Chapter 1
Basic Phenomena and Applications to Thin Films

1.1 Surface Tension at Fluid Interfaces

Common liquids and gases under normal conditions can have different densities and other physical properties. However, the word "fluid" is used to describe all of them because of one common property: they can deform easily under the action of even small forces. The spatial distribution of molecules in fluids lacks order seen in crystalline solids, which tend to resist deformation elastically. While in principle motion of a fluid can be analyzed by describing motion of individual molecules and interactions between them, in practice this is impossible because of the very large number of molecules involved. Even a tiny spherical water drop of radius 1 mm contains on the order of 10^{20} molecules. Fortunately, it is the large number of molecules that allows one to drastically simplify the mathematical description of fluids by using the so-called continuum hypothesis, which is the core of all models described in this book. Instead of tracking individual molecules, we consider fluid elements, each containing a large number of molecules and characterized by its velocity of motion and pressure. Fluid elements are assumed to be small compared to the distances over which velocity, pressure, and all physical properties change. By applying the principle of conservation of mass and Newton's second law to fluid elements, the classical continuity and Navier–Stokes equations for velocity and pressure fields are derived, as discussed in a number of fluid mechanics textbooks, e.g., in Acheson [1] and Batchelor [12].

When several fluids are present, their molecules can be either mixed together or separated by surfaces which we refer to as fluid interfaces. For example, the liquid–gas interface which defines a bubble in a liquid is a fluid interface. In order to be able to specify the correct boundary conditions at fluid interfaces, it is important to understand which physical phenomena have to be introduced in the continuum description of fluids to account for the presence of interfaces.

Consider a small segment of an interface between two different fluids, e.g., water and air. Suppose we can zoom into this region to observe individual molecules making up the two fluids: then the snapshot of the region near the interface might

V.S. Ajaev, *Interfacial Fluid Mechanics: A Mathematical Modeling Approach*,
DOI 10.1007/978-1-4614-1341-7_1, © Springer Science+Business Media, LLC 2012

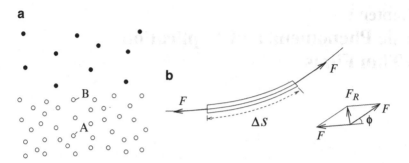

Fig. 1.1 (**a**) An illustration of the molecular basis of the surface tension: a zoom-in of an interfacial region; (**b**) Sketch for the derivation of the formula for the pressure jump at a liquid–gas interface

look like the sketch in Fig. 1.1a. Molecules of the upper and lower fluids are represented schematically by the filled and empty circles, respectively. Molecules interact with each other, but the interaction force is known to decrease rapidly with distance, so each molecule has measurable interaction only with the ones which are relatively close. This implies, e.g., that a molecule marked "A" in Fig. 1.1a does not feel the presence of any of the molecules of the upper fluid. A molecule marked "B", on the other hand, feels the presence of the upper fluid and interacts with fewer molecules since the upper fluid is less dense. When the area of the interface increases, the percentage of molecules experiencing less interaction, such as the molecule marked "B", increases, meaning that work is done against the intermolecular forces. Thus, interfaces tend to resist stretching. The argument presented here is valid for molecules of different types, as in Fig. 1.1a or the same type but having different density (e.g., water and its own vapor).

In reality, interfaces between liquid and gas are two-dimensional surfaces in the three-dimensional space, but a cross-section of an interface in an arbitrary plane is a smooth planar curve, as illustrated in Fig. 1.1b. For simplicity, we start by assuming that the shape of the interface does not change in the direction normal to the plane of the sketch; such configurations are referred to as two-dimensional. We parameterize the curve representing the interface by the arclength variable s and consider a control volume shown in Fig. 1.1b, i.e., a thin layer containing the segment of the interface between s_0 and $s_0 + \Delta s$. The dimension of the control volume in the direction normal to the plane is assumed equal to l. The tendency of interfaces to resist stretching means that there are two tension forces acting on the control volume from the sides in the plane of the sketch, both of the same magnitude F but acting in different directions, as shown in Fig. 1.1b. The vector sum of these forces gives a component in the direction normal to the interface, of absolute value $F_R = 2F \sin(\phi/2)$, where ϕ is the angle shown in the sketch. This angle is small for $\Delta s \ll 1$, so we can write $F_R \approx F\phi \approx F\phi'(s_0)\Delta s$. The last step is based on the Taylor expansion near $s = s_0$.

By applying Newton's second law to the control volume shown in the sketch and taking the limit of the layer thickness (and thus the mass and momentum of the liquid inside the control volume) approaching zero, we conclude that the vector sum of all forces acting on the control volume has to be zero. Considering the direction locally normal to the interface, this implies that the force F_R has to be balanced by the difference of forces acting from the liquid side and the gas side, which are expressed in terms of the local values of pressure in the liquid (p) and the gas (p_g), leading to the balance of the form

$$F\phi'(s_0)\Delta s = (p_g - p)l\Delta s. \tag{1.1}$$

In the arclength representation of a curve, the curvature is equal to the rate of change of the slope angle, so $\phi'(s_0)$ can be replaced by the curvature κ, leading to the desired equation

$$p - p_g = -\sigma\kappa. \tag{1.2}$$

This equation simply states that at a curved interface there is a pressure jump proportional to the local curvature. The coefficient of proportionality, σ, is called the surface tension. It is related to the previously introduced tension force F acting on the control volume by $\sigma = F/l$. Equation (1.2) is the two-dimensional version of the classical Young–Laplace equation.

Surface tension is essential in explaining many physical phenomena, i.e., water climbing up small tubes placed in fluid containers, and difficulties in extracting liquid from small pipes. The ability of porous media to hold large amounts of liquid is due to the significance of surface tension in small pores. Many biological processes, such as oxygen exchange in the lungs, depend on the surface tension force. Small insects such as the water strider can walk on water using the surface tension forces, which are significant enough to compensate their weight.

1.2 Capillary Statics

Static shapes of fluid interfaces are determined by the combined action of surface tension forces and gravity. If a solid boundary is present, the interface shape will also depend on the physical properties of this boundary. The effect of solid walls on static interface shapes can be easily observed in a glass partially filled with water. While flat away from the wall of the glass, the air–water interface is usually curved near the so-called contact line, i.e., the line along which the water surface meets the solid wall. Consider a two-dimensional version of this configuration, shown in Fig. 1.2a, with the air–liquid interface shape represented by the thick curve $y = h(x)$ and the contact line corresponding to the point $(0, h_{\max})$ at which this curve intersects with the vertical solid boundary. We use Cartesian coordinates such that the y-axis is along the vertical solid wall, the x-axis is horizontal, and $y = 0$ corresponds to the nearly flat interface shape far away from the wall (we assume that the liquid is in a container large enough so that the interface flattens out as x is increased).

Consider a force balance for a control volume which is a region containing all three phases, with x between $-\Delta x/2$ and $\Delta x/2$ and y between $h_{\max} - \Delta y/2$ and $h_{\max} + \Delta y/2$, as shown in Fig. 1.2a. As in the previous section, we assume that the

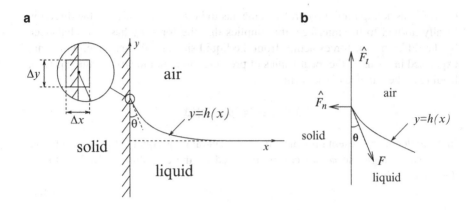

Fig. 1.2 (a) Sketch of an air–liquid interface near a vertical solid wall and an enlarged view of the contact-line region showing the control volume used in the derivation of the force balance at the contact line. (b) A diagram showing forces acting on the control volume enclosing the contact line

length of the control volume in the direction normal to the sketch is l. In the limit of both Δx and Δy approaching zero (but still in the framework of the continuum approximation), Newton's second law dictates that the sum of all forces acting on the control volume is zero. The force $F = \sigma l$ due to surface tension at the air–liquid interface does not vanish in this limit and therefore has to be balanced by a force acting on the control volume at the solid wall. The latter force can be decomposed into the component $\hat{F}_n$ normal to the wall and the component $\hat{F}_t$ along the wall, as shown in Fig. 1.2b. The elastic deformation of the solid is required to balance the normal force $\hat{F}_n$, but the situation with the tangential component is more complicated. To clarify the origin of $\hat{F}_t$, we note that the solid–air and solid–liquid interfaces seen in Fig. 1.2 are characterized by different energies per unit area, denoted by σ_{sa} and σ_{sl}, respectively. The difference is due to the fact that the energies of intermolecular interactions at solid–fluid interfaces depend on the density and the chemical composition of the fluid and therefore are not the same for the solid–air and solid–liquid interfaces. When the contact line is displaced by a small amount δh, the energy change at the solid–fluid interface is given by $(\sigma_{sl} - \sigma_{sa})l\delta h$, which can be interpreted as work done against the tangential force, leading to the expression

$$\hat{F}_t = (\sigma_{sa} - \sigma_{sl})l. \tag{1.3}$$

The condition of the zero vector sum of all forces shown in Fig. 1.2b implies that their projections onto the vertical line $x = 0$ have to add up to zero, leading to the classical Young's equation,

$$\sigma \cos \theta = \sigma_{sa} - \sigma_{sl}. \tag{1.4}$$

Here θ is the angle between the local tangent to the air–liquid interface and the solid wall, known as the *contact angle*. Equation (1.4) determines a unique static value of the contact angle for any pair of solid and liquid (assuming the gas phase is air under normal conditions) and thus allows one to write a boundary condition for the interface shape $y = h(x)$ in the form

$$h_x(0) = -\cot\theta. \tag{1.5}$$

It is important to emphasize that when a small displacement of the contact line is discussed in the above derivation, it does not imply creating new segments of solid–fluid interfaces by bringing atoms from the bulk of the crystalline solid, so the physical interpretation of the surface energies σ_{sa} and σ_{sl} is not quite the same as that of the surface tension at an easily deformable fluid interface, discussed in the previous section.

The surface energies of the solid–liquid and solid–air interfaces depend not only on the chemical compositions of the solid and fluids involved but also on how smooth the solid surface is, so for example polishing the surface can affect the contact angle. In addition, even slight contamination of interfaces can result in changes of the contact angle. It is interesting to note that the contact angle for pure water on very clean smooth glass surface is approximately zero, which is in obvious contrast with observations of rain droplets on car windshields.

Even though Young's equation is widely used in modeling of contact lines, its experimental verification is rather limited due to lack of independent measurements of the solid–liquid and solid–air surface energies [70]. It is also important to note that the equation (1.3) for the tangential force, $\hat{F}_t$, is based on the assumption that the elastic stresses in the solid affect only the normal force component at the solid–fluid interface, $\hat{F}_n$.

The air–liquid interface shape illustrated in Fig. 1.2a is flat away from the wall and satisfies the contact angle condition (1.5) at $x = 0$. Aside from the special case of $\theta = \pi/2$, these two conditions can be satisfied only when the interface is curved in a region near the wall. To determine the exact shape of the interface in this region, we employ (1.2), which relates the pressure jump at the interface and its local curvature. Since the gas phase in Fig. 1.2a is air at constant atmospheric pressure p_0, (1.2) takes the form

$$p - p_0 = -\sigma\kappa. \tag{1.6}$$

The curvature κ can be expressed in terms of the derivatives of the yet unknown interface shape $y = h(x)$ as:

$$\kappa = \frac{h_{xx}}{(1 + h_x^2)^{3/2}}. \tag{1.7}$$

By applying (1.6) to the flat portion of the interface, we observe that the liquid pressure p is equal to p_0 there. In the curved region of the interface, the value of p deviates from p_0 by the hydrostatic pressure difference, so (1.6) can be written in the form

$$\sigma h_{xx} \left(1 + h_x^2\right)^{-3/2} = \rho g h, \tag{1.8}$$

where ρ is the density of the liquid, g is the acceleration of gravity. Multiplying both sides of (1.8) by h_x and integrating once gives

$$\left(1+h_x^2\right)^{-1/2} = c_1 - \frac{h^2}{2a^2}. \tag{1.9}$$

Here we introduced the so-called *capillary length*, $a = \sqrt{\sigma/\rho g}$. The condition of a flat interface at $x = \infty$ requires that $c_1 = 1$. With this value, (1.9) can be used to determine the maximum interface height h_{max}, which in the configuration of Fig. 1.2a is reached at $x = 0$. The interface slope at $x = 0$ is defined by (1.5). After substituting the values of c_1 and the slope into (1.9) we find

$$h_{max} = \sqrt{2}a(1 - \sin\theta)^{1/2}. \tag{1.10}$$

If the contact angle θ is zero (a situation described as "perfect" or "complete" wetting), h_{max} reaches the highest possible value of $\sqrt{2}a$.

Equation (1.9) with $c_1 = 1$ can be written as:

$$dx = \left(\frac{h^2}{2a^2} - 1\right) \frac{dh}{\frac{h}{a}\sqrt{1 - \frac{h^2}{4a^2}}} \tag{1.11}$$

and then integrated using standard substitutions, giving an implicit formula for the shape of the interface,

$$x = a \operatorname{arccosh} \frac{2a}{h} - 2a\left(1 - \frac{h^2}{4a^2}\right)^{1/2} + c_2. \tag{1.12}$$

The constant c_2 is found from the condition $h(0) = h_{max}$ and the equation (1.10): $c_2 = \sqrt{2}a(1 + \sin\theta)^{1/2} - a\operatorname{arccosh}\left[\sqrt{2}(1 - \sin\theta)^{-1/2}\right]$. Note that the value of the contact angle θ enters the solution only through the constant c_2, so solutions for different contact angles are in fact parts of the same curve but shifted in the horizontal direction.

Our derivation of (1.12) is based on the assumption that the container is large enough so that the liquid surface flattens out away from the wall. The definition of "large" in this context is based on comparison between the characteristic cross-sectional size of the container L and the capillary length a (equal to 2.7 mm for the air–water interface with the surface tension taken to be $\sigma = 0.073$ N/m). The shapes of the interface are different depending on the ratio L/a: from essentially flat everywhere except near the walls (large L/a) to nearly constant curvature (small L/a). The latter case is described by (1.6) in the limit of negligible gravity, equivalent to $\kappa = \text{const}$. The quantity $(L/a)^2$ is referred to as the Bond number (denoted by "Bo") and is a measure of the importance of gravity relative to surface tension forces. Using our expression for a, the definition of the Bond number can be written in the following standard form:

$$\text{Bo} = \frac{\rho g L^2}{\sigma}. \tag{1.13}$$

In microscale applications the Bond number is typically small, indicating that the gravity has relatively little effect on the shapes of fluid interfaces, which are dominated by the capillary forces.

1.3 Liquid Transport on Microscale

A significant amount of current research in interfacial fluid mechanics is motivated by applications in the field of microfluidics, which usually refers to the study of fluid flow in artificial microsystems. Several books have been published on the subject of microfluidics in recent years [16, 25, 33, 77, 127]. Understanding liquid transport is essential for designing microfluidic devices in which liquid flow is used to deliver reagents to the site of a chemical reaction and for a variety of other tasks. Many of our everyday notions about efficient means to transport liquid fail when applied to microscale flows. Most notably, from the everyday experience of pouring tea or coffee into a cup to large-scale applications such as water towers and hydroelectric power plants, gravity plays an important role in transporting liquids through hydrostatic pressure differences. However, it turns out that gravity is not an efficient method for transporting liquid on the microscale and can often be assumed negligible in mathematical models of microscale flows.

In order to better understand flows under the action of gravity and estimate the rate at which they can transport liquid, let us consider viscous flow in a thin film down an inclined plane, sketched in Fig. 1.3. We note that a vertical wall is a special case of this configuration when the inclination angle α is $\pi/2$, while a horizontal surface corresponds to $\alpha = 0$. The film is assumed to be of uniform thickness d, so we ignore possible formation of ripples on its surface, which turns out to be a reasonable assumption for sufficiently thin slowly moving films of viscous liquid. The Cartesian coordinates are chosen such that the x-axis is directed along the solid surface and the y-axis is normal to it. We assume that the flow is steady

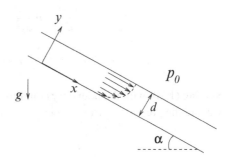

Fig. 1.3 Gravity-driven flow of a thin liquid film down an incline

and two-dimensional, meaning that the velocity distribution is independent of time
and also does not change in the direction of z (normal to the plane of the sketch
of Fig. 1.3).

Steady flow of a Newtonian incompressible fluid of density ρ and viscosity μ is
governed by the Navier–Stokes and continuity equations [1, 12] in the form

$$u\frac{\partial u}{\partial x} + v\frac{\partial u}{\partial y} = -\frac{1}{\rho}\frac{\partial p}{\partial x} + \nu\left(\frac{\partial^2 u}{\partial x^2} + \frac{\partial^2 u}{\partial y^2}\right) + g\sin\alpha, \tag{1.14}$$

$$u\frac{\partial v}{\partial x} + v\frac{\partial v}{\partial y} = -\frac{1}{\rho}\frac{\partial p}{\partial y} + \nu\left(\frac{\partial^2 v}{\partial x^2} + \frac{\partial^2 v}{\partial y^2}\right) - g\cos\alpha, \tag{1.15}$$

$$\frac{\partial u}{\partial x} + \frac{\partial v}{\partial y} = 0. \tag{1.16}$$

Here u and v are the velocity components in the directions of the x- and y-
axes, respectively, p is the pressure, $\nu = \mu/\rho$. The standard boundary conditions
(discussed, e.g., in Acheson [1]) at the solid wall and the liquid–air interface are

$$u = v = 0 \quad \text{at} \quad y = 0, \tag{1.17}$$

$$p = p_0, \quad \frac{\partial u}{\partial y} = 0 \quad \text{at} \quad y = d. \tag{1.18}$$

Equation (1.17) expresses the fact that liquid does not slip (i.e., does not have a
nonzero velocity) along the stationary solid surface and does not penetrate into the
solid. According to (1.18), there is no jump in the pressure or tangential stress at the
liquid–air interface; air is at the atmospheric pressure p_0 and the tangential stresses
in air are assumed negligible since the viscosity of air is much smaller than that
of the liquid. Note that the surface tension discussed in the previous two sections
does not appear in the boundary condition for pressure here because the liquid–air
interface is flat.

If, motivated by constant value of the film thickness, we assume that u is not a
function of x, then using (1.16) and (1.17) we immediately determine that v is zero
everywhere. The Navier–Stokes equations (1.14)–(1.15) then simplify to

$$-\frac{\partial p}{\partial x} + \mu\frac{d^2 u}{dy^2} + \rho g\sin\alpha = 0, \tag{1.19}$$

$$-\frac{\partial p}{\partial y} - \rho g\cos\alpha = 0. \tag{1.20}$$

Solving (1.20) with the boundary condition for pressure at the interface from (1.18)
we find the pressure distribution in the form

$$p = -\rho g(y - d)\cos\alpha + p_0. \tag{1.21}$$

Thus, the pressure is not a function of x and therefore (1.19) becomes

$$\mu \frac{d^2 u}{dy^2} + \rho g \sin \alpha = 0. \tag{1.22}$$

Integrating this ordinary differential equation twice we obtain

$$u = -\frac{\rho g \sin \alpha}{2\mu} y^2 + c_1 y + c_2. \tag{1.23}$$

The no-slip condition at the solid boundary from (1.17) requires that c_2 is set to zero. The remaining constant is found using the tangential stress condition from (1.18), leading to

$$u = \frac{\rho g \sin \alpha}{2\mu} (2yd - y^2). \tag{1.24}$$

This velocity profile is illustrated by arrows in Fig. 1.3. The maximum velocity is reached at the interface ($y = d$) and is equal to $\rho g d^2 \sin \alpha / 2\mu$. A useful measure of how efficiently liquid is transported is the average velocity,

$$\bar{u} \equiv \frac{1}{d} \int_0^d u \, dy = \frac{\rho g d^2 \sin \alpha}{3\mu}. \tag{1.25}$$

For a film of water of thickness of $d = 10 \, \mu$m on a vertical solid surface ($\alpha = \pi/2$), the average velocity is on the order of 0.1 mm/s. This value is significantly lower than needed for typical microfluidic applications. The average velocity is even smaller for surfaces inclined at an angle $\alpha < \pi/2$, not to mention that the dependence of flow on the orientation of the surface is something to be avoided in portable microfluidic devices. Thus, we conclude that gravity-driven flow is not an effective mechanism for transporting liquids in small-scale devices.

Gravity is not the only common mechanism used for transporting fluids on the macroscale. For example, when air is pumped into car or bicycle tires, it is transported by the differences in fluid pressure, not by gravity. The same principle is used in many large-scale industrial applications. Let us investigate if the same approach can be used for small-scale systems. Consider two liquid reservoirs maintained at different pressures and connected by a long channel of height $2d$, as shown in Fig. 1.4. We consider the two-dimensional version of the problem, i.e., assume that pressure and velocity components are functions of x and y (shown in the sketch) and not z. One can think about our solution as a limiting case of flow in a channel of rectangular cross-section with one side much larger than the other.

The Navier–Stokes equations for this flow can be simplified using the same assumptions (steady flow, $u = u(y)$) as in the derivation of equations (1.19)–(1.20) for the gravity-driven flow, resulting in

Fig. 1.4 Sketch of a channel connecting two reservoirs of liquid at different pressures

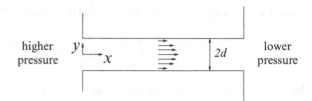

$$-\frac{\partial p}{\partial x} + \mu \frac{d^2 u}{dy^2} = 0, \qquad (1.26)$$

$$-\frac{\partial p}{\partial y} - \rho g = 0. \qquad (1.27)$$

Integrating (1.27) we find that $p = -\rho g y + f(x)$. According to (1.26), f is a linear function of x. It is convenient to introduce a horizontal pressure gradient $G = |f'(x)|$, equal to the pressure difference between the reservoirs divided by the length of the channel. Then the horizontal velocity satisfies

$$\frac{d^2 u}{dy^2} = -\frac{G}{\mu} \qquad (1.28)$$

with the no-slip conditions at the top and bottom walls,

$$u = 0 \quad \text{at} \quad y = \pm d. \qquad (1.29)$$

The solution of (1.28)–(1.29) is

$$u = \frac{G}{2\mu} \left(d^2 - y^2 \right), \qquad (1.30)$$

shown by arrows in Fig. 1.4. Since the velocity profile is symmetric, the average value can be calculated from

$$\bar{u} = \frac{1}{d} \int_0^d u \, dy = \frac{G d^2}{3\mu}. \qquad (1.31)$$

If a pressure difference of 10^4 Pa is applied to a channel of length 1 cm and height 20 μm, the average velocity of water flow found from (1.31) is approximately 3 cm/s, significantly higher than the previously discussed velocity estimate for gravity-driven flows. Also, in contrast to the acceleration of gravity g, the pressure gradient can be increased by making changes in the design of the experimental apparatus. These advantages of pressure-driven flows led to their successful use in many experiments with liquids on the microscale. However, it is important to remember that while reaching high values of the pressure gradient is relatively easy in a research lab environment, it turns out to be a challenge in small-scale devices for commercial applications, especially since many such devices have to be portable. Furthermore, in lab-on-a-chip devices liquids have to be manipulated in complicated networks of microchannels, not just moved from one reservoir to another [120]. Another difficulty stems from the fact that the parabolic velocity profile illustrated by arrows in Fig. 1.4 implies that, e.g., a drop or a biological cell transported by such liquid flow can have very different velocity depending on how far it is from the channel walls. Thus, we conclude that even though pressure-driven flow can be a reasonably efficient mechanism for transporting liquids on microscale, several difficulties are encountered in its practical use for microfluidic applications. These difficulties motivated research aimed at finding new approaches to liquid transport on the microscale. A significant part of this research focused on better understanding of the physical properties of the liquid–gas, liquid–liquid, and solid–liquid interfaces and resulted in development of many mathematical models discussed in this book. In the following section we provide an illustration by discussing how an observation of surface tension dependence on temperature resulted in a new approach for generating flow in very thin liquid films and discuss its practical applications.

1.4 Thermocapillary Flow in a Thin Liquid Film

1.4.1 Introduction to Thermocapillary Phenomena

Experimental studies indicate that the surface tension at a fluid interface depends on its local temperature. The solid line in Fig. 1.5a illustrates a typical shape of the measured curve for surface tension of water versus temperature. For many liquids

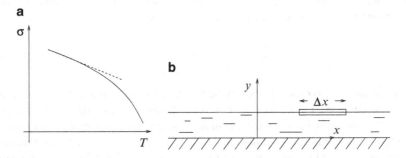

Fig. 1.5 (**a**) Sketch of the surface tension as a function of temperature for water over a range of temperatures relevant for most practical applications; *dashed line* shows the linear approximation used in many models; (**b**) Sketch of a uniform thin liquid film on a horizontal surface and Cartesian coordinates; the temperature gradient is in the *x*-direction

and temperature differences under $\sim$100 K surface tension can be approximated by
the formula

$$\sigma = \sigma_0 - \gamma(T - T_0), \tag{1.32}$$

where σ_0 is the value of surface tension measured at a reference temperature T_0 and
γ is a positive number ($\gamma = 1.5 \times 10^{-4}$ N/(m K) for air–water interface at 20°C).
Equation (1.32) corresponds to the dashed line in Fig. 1.5a.

The result of the dependence of the surface tension on the local temperature
is that a so-called thermocapillary (or "Marangoni") flow can be generated by
imposing a temperature gradient along the interface. The term "Marangoni flow" is
also often used in a broader sense for describing any flow generated by gradients of
surface tension. Such gradients can be present even under the isothermal conditions,
as discussed in detail in Chap. 7. To understand how thermocapillary flow can be
generated consider a thin film of uniform thickness d on a flat substrate shown in
Fig. 1.5b. Also shown in the sketch is a control volume enclosing a small segment
of the interface of length Δx in the x-direction and of length l in the direction
normal to the sketch. As long as the interface is flat, the projections of surface
tension forces on the vertical direction are zero, so there is no pressure jump
across the interface. However, if the surface tension σ is a function of x due to
temperature variations along the interface, then the vector sum of the surface tension
forces acting on the control volume has a non-zero projection onto the x-direction,
$\Delta F = (\sigma(x+\Delta x) - \sigma(x))l$. Here we use the same two-dimensional framework as in
the previous sections, assuming no variation of the surface tension in the direction
normal to the plane of the sketch in Fig. 1.5b. From Newton's second law, the sum
of all forces acting on the control volume has to be zero, which means that ΔF has
to be balanced by the tangential force at the interface due to viscous friction in the
liquid,

$$\mu \frac{\partial u}{\partial y} \Delta x l, \tag{1.33}$$

where u is the velocity component in the x-direction. As in Sect. 1.3, the tangential
stress from the air side is neglected. In the limit of $\Delta x \to 0$, the force balance in the
x-direction leads to the boundary condition at the interface,

$$\mu \frac{\partial u}{\partial y} = \frac{d\sigma}{dx} \quad \text{at} \quad y = d. \tag{1.34}$$

In general, there is no reason to expect the interface to remain flat as the thermocapillary flow develops, but it turns out that for a special case of a linear temperature
profile $T(x)$ along the interface there is in fact a solution corresponding to a uniform
flat interface and unidirectional flow in the film, described by $u = u(y)$. To find this
velocity profile, we observe that equation (1.19) for a horizontal surface and zero
pressure jump across the liquid–air interface reduces to

$$\frac{d^2 u}{dy^2} = 0. \tag{1.35}$$

Using the linear dependence of the surface tension on temperature from (1.32) we can rewrite (1.34) as:

$$\frac{du}{dy} = -\frac{\gamma}{\mu}\frac{dT}{dx} \quad \text{at} \quad y = d. \tag{1.36}$$

Solving (1.35) with the no-slip condition at the solid wall, $u(0) = 0$, and the condition (1.36) at the liquid–air interface results in

$$u = -\frac{\gamma}{\mu}\frac{dT}{dx}y. \tag{1.37}$$

Positive values of the horizontal velocity correspond to negative temperature gradients. In general, interfacial temperature gradients in common liquids result in liquid flow from hot to cold regions. The average velocity of the thermocapillary flow described by (1.37) is

$$\bar{u} = \frac{\gamma d}{2\mu}\left|\frac{dT}{dx}\right|. \tag{1.38}$$

Even with a moderate temperature gradient of $\sim$1 K/cm the average velocity is much higher than that of the gravity-driven flow on a vertical plane, found from (1.25). Thus, thermocapillary flow can be a promising mechanism for transporting liquids at small-scales.

1.4.2 Modeling of a Thin Film with Deforming Surface

In practical applications of thermocapillary flows, the interface can deform under the action of temperature gradients. To illustrate this, let us consider a liquid film of initially uniform thickness d and use the same two-dimensional framework as before (i.e., assume no z-dependence). However, instead of assuming the temperature at the interface T^i to be a linear function of x let us use the following temperature distribution

$$T^i = T_0 + \Delta T e^{-\alpha x^2}. \tag{1.39}$$

In practice, the temperature profile will be changing in time and has to be found by solving a system of equations for *both* liquid flow and heat transfer. However, in an effort to focus on the description of liquid flow in the film, we assume that the temperature profile along the interface is well approximated by (1.39) and does not change with time. According to (1.39), the temperature decays from the maximum value $T_0 + \Delta T$ to the environment temperature T_0 over the distance of the order of $\alpha^{-1/2}$. The value of $\alpha^{-1/2}$ can be used as a *characteristic length* of change of the interfacial temperature, meaning an order-of-magnitude estimate of the distance over which the temperature changes by the amount ΔT.

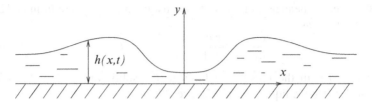

Fig. 1.6 A snapshot of the liquid–air interface deforming under the action of thermocapillary stresses; the maximum of the interfacial temperature corresponds to $x = 0$

The notion of the characteristic length can be used to estimate typical values of derivatives of a physical quantity without doing any calculations, by writing e.g.,

$$\left| \frac{dT^i}{dx} \right| \approx \frac{\Delta T}{\alpha^{-1/2}}, \qquad \left| \frac{d^2 T^i}{dx^2} \right| \approx \frac{\Delta T}{\alpha^{-1}}, \qquad \text{etc.} \tag{1.40}$$

The derivatives of T^i can, of course, be calculated exactly from (1.39). For example, $|dT^i/dx|$ turns out to vary between 0 and $\Delta T \sqrt{2\alpha} e^{-1/2}$. However, the importance of estimates such as (1.40) is that they can be made even for physical quantities whose exact dependence on the spatial variable is unknown, as long as the characteristic length of their variation can be estimated. Note that even though (1.39) allows one to obtain more accurate values for the characteristic length of change of temperature, using $\alpha^{-1/2}$ is equally justified as long as one is only interested in order-of-magnitude estimates.

In contrast to the previous examples of flows in thin liquid films, the film thickness here is not expected to remain uniform. As the thermocapillary flow of liquid away from the center of the heated area develops, the interface will deform and the film will gradually be depleted near $x = 0$, resulting in interface shapes such as the one illustrated in Fig. 1.6. The Navier–Stokes and continuity equations for the two-dimensional flow in the film are

$$\frac{\partial u}{\partial t} + u \frac{\partial u}{\partial x} + v \frac{\partial u}{\partial y} = -\frac{1}{\rho} \frac{\partial p}{\partial x} + \nu \left(\frac{\partial^2 u}{\partial x^2} + \frac{\partial^2 u}{\partial y^2} \right), \tag{1.41}$$

$$\frac{\partial v}{\partial t} + u \frac{\partial v}{\partial x} + v \frac{\partial v}{\partial y} = -\frac{1}{\rho} \frac{\partial p}{\partial y} + \nu \left(\frac{\partial^2 v}{\partial x^2} + \frac{\partial^2 v}{\partial y^2} \right) - g, \tag{1.42}$$

$$\frac{\partial u}{\partial x} + \frac{\partial v}{\partial y} = 0. \tag{1.43}$$

Since the flow is generated due to the temperature gradient at the interface, the characteristic length of change of u and v in the x-direction is the same as that of T^i. As discussed above, this length is of the order of $\alpha^{-1/2}$. However, to make the following discussion applicable to a wider range of thin film flows, we avoid making direct references to the particular temperature profile given by (1.39) and therefore use a general notation, L_x, for the characteristic length scale of the flow in the horizontal direction. In order to satisfy the standard boundary conditions (1.17) at the solid surface the velocity components have to change in the y-direction over the scale of the film thickness, on the order of the initial thickness d. If the ratio of the scales, $\varepsilon = d/L_x$, is small then

$$\left|\frac{\partial^2 u}{\partial x^2}\right| \ll \left|\frac{\partial^2 u}{\partial y^2}\right|, \qquad \left|\frac{\partial^2 v}{\partial x^2}\right| \ll \left|\frac{\partial^2 v}{\partial y^2}\right|, \tag{1.44}$$

and therefore the second derivatives with respect to x in (1.41) and (1.42) can be neglected.

Suppose U is the characteristic value of u, the horizontal velocity component. If we assume that v is of the same order as U, then the equation of continuity together with (1.17) will immediately result in $v = 0$ and therefore a nonphysical prediction of constant film thickness. The contradiction can be avoided only if v is of the order of εU so that both terms in (1.43) are equally important. With this assumption, the terms on the left-hand-side of (1.41) can be neglected as long as the Reynolds number, Re$= Ud/\nu$, is not large. To justify this, we make the following estimates:

$$\frac{|u_t + uu_x + vu_y|}{\nu|u_{yy}|} \sim \frac{U^2/L_x}{\nu U/d^2} = \varepsilon \text{Re}. \tag{1.45}$$

Here we estimated the characteristic time of change of U to be L_x/U. Similar arguments can be used to estimate the left-hand-side terms of (1.42). Thus, for $\varepsilon \ll 1$ and $\varepsilon \text{Re} \ll 1$ the system of governing equations, (1.41)–(1.43), simplifies to

$$-\frac{\partial p}{\partial x} + \mu \frac{\partial^2 u}{\partial y^2} = 0, \tag{1.46}$$

$$-\frac{\partial p}{\partial y} - \rho g = 0, \tag{1.47}$$

$$\frac{\partial u}{\partial x} + \frac{\partial v}{\partial y} = 0. \tag{1.48}$$

The term $\mu \partial^2 v/\partial y^2$ in the second equation is omitted based on the following argument. If this term is retained, then instead of (1.47) one has to consider

$$-\frac{\partial p}{\partial y} + \mu \frac{\partial^2 v}{\partial y^2} - \rho g = 0 \tag{1.49}$$

or, after integration in y,

$$p = \mu \frac{\partial v}{\partial y} - \rho g y + f(x). \tag{1.50}$$

Here we introduced a new function $f(x)$ which can be found by substituting (1.50) into (1.46). Since the objective here is to estimate various terms contributing to the pressure p, we simply observe that based on (1.46), the magnitude of $f(x)$ is of the order of $\mu U/\varepsilon d$. The contribution of the first term on the right-hand-side of (1.50) is estimated to be $\sim \mu \varepsilon U/d$ and can therefore be neglected in the limit of small ε. Thus, using (1.47) instead of (1.49) is justified.

Equations (1.46)–(1.48) are the core of the so-called *lubrication-type approach* to modeling of thin liquid films. The origin of the name has to do with the classical lubrication theory of Reynolds which deals with a layer of liquid between two solid surfaces. However, the problem of interest here is different since only one boundary is a solid surface, while the other one is a deforming fluid interface.

Let us now discuss the boundary conditions at the air–liquid interface, defined by a time-dependent thickness of the film, $y = h(x,t)$. In this section we simply state these boundary conditions in the limit of small ε; their derivation is provided in Sect. 1.5*. The first interfacial boundary condition is the so-called kinematic condition which relates the rate of change of the interface height and the components of the local flow velocity,

$$\frac{\partial h}{\partial t} + u\frac{\partial h}{\partial x} - v = 0 \quad \text{at} \quad y = h(x,t). \tag{1.51}$$

The other two conditions reflect the balances of normal and tangential components of the stress and are written in the form

$$p - p_0 = -\sigma\frac{\partial^2 h}{\partial x^2} \quad \text{at} \quad y = h(x,t), \tag{1.52}$$

$$\mu\frac{\partial u}{\partial y} = \frac{\partial \sigma}{\partial x} \quad \text{at} \quad y = h(x,t). \tag{1.53}$$

Based on (1.32), the tangential stress condition (1.53) can be written in terms of the interfacial temperature profile T^i,

$$\frac{\partial u}{\partial y} = -\frac{\gamma}{\mu}\frac{dT^i}{dx} \quad \text{at} \quad y = h(x,t). \tag{1.54}$$

The standard boundary conditions at the solid surface, (1.17), complete our formulation of the problem.

1.4.3 Nondimensional Variables and Evolution Equation

The problem formulated in the previous subsection involves a number of dimensional parameters such as liquid viscosity and surface tension. As is often the case in fluid mechanics, the actual solution of the problem turns out to depend only on fewer dimensionless combinations of the dimensional quantities. These combinations are easily identified when the problem is reformulated in terms of nondimensional variables defined by

$$(\tilde{y}, \tilde{h}) = \frac{(y, h)}{d}, \quad \tilde{x} = \frac{x}{L_x}, \quad \tilde{t} = \frac{t}{L_x/U}, \quad \tilde{u} = \frac{u}{U}, \quad \tilde{v} = \frac{v}{\varepsilon U},$$

$$\tilde{p} = \frac{p - p_0}{P}, \quad \tilde{T}^i = \frac{T^i - T_0}{\Delta T}. \tag{1.55}$$

Since the flow is caused by the tangential stress at the interface, the characteristic velocity $U = \gamma \Delta T d / \mu L_x$ is found from the tangential stress balance, (1.54), together with the formula (1.39) for the interfacial temperature profile. The choice of the characteristic pressure difference, $P = \sigma_0 d / L_x^2$, reflects the fact that pressure gradients in the film appear because of the deformation of the interface and are therefore directly related to curvature, as expressed by (1.52). When a problem is formulated in nondimensional terms, it is common to state that variables are scaled by their characteristic values. For example, $\tilde{y}$ can be referred to as the vertical coordinate scaled by d.

The governing equations (1.46)–(1.48) take the following nondimensional form:

$$\frac{\partial \tilde{p}}{\partial \tilde{x}} = \varepsilon^{-3} \text{Ca} \frac{\partial^2 \tilde{u}}{\partial \tilde{y}^2}, \tag{1.56}$$

$$\frac{\partial \tilde{p}}{\partial \tilde{y}} + \text{Bo} = 0, \tag{1.57}$$

$$\frac{\partial \tilde{u}}{\partial \tilde{x}} + \frac{\partial \tilde{v}}{\partial \tilde{y}} = 0. \tag{1.58}$$

Here we use the Bond number first introduced at the end of Sect. 1.2 (and written as $\text{Bo} = \rho g L_x^2 / \sigma_0$ in the notation of the present section) and introduce the *capillary number*,

$$\text{Ca} = \frac{\mu U}{\sigma_0}. \tag{1.59}$$

The capillary number measures the importance of viscous effects relative to surface tension and is typically small. For example, when water is at 20°C ($\mu = 10^{-3} \, \text{N s m}^{-2}$) and in contact with air ($\sigma_0 = 0.073 \, \text{N m}^{-1}$), even for a large flow velocity of 1 m/s the capillary number is 0.014. However, in (1.56) the capillary number multiplies a large parameter ε^{-3}, so in a thin film both surface tension and viscous effects can be important. In order to include both effects in the mathematical model, ε^3 and Ca have to be of the same order. This can be achieved by simply defining L_x to be $d \, \text{Ca}^{-1/3}$ (recall that $\varepsilon = d/L_x$). Here we also assume that $d \, \text{Ca}^{-1/3}$ and the characteristic length $\alpha^{-1/2}$ imposed by the temperature profile from (1.39) are of the same order. Equation (1.56) then takes the simple form

$$\frac{\partial \tilde{p}}{\partial \tilde{x}} = \frac{\partial^2 \tilde{u}}{\partial \tilde{y}^2}. \tag{1.60}$$

The interfacial boundary conditions (1.51), (1.52), and (1.54) are written in terms of the nondimensional variables as:

$$\frac{\partial \tilde{h}}{\partial \tilde{t}} + \tilde{u}\frac{\partial \tilde{h}}{\partial \tilde{x}} - \tilde{v} = 0 \quad \text{at} \quad \tilde{y} = \tilde{h}(\tilde{x},\tilde{t}), \tag{1.61}$$

$$\tilde{p} = -\frac{\partial^2 \tilde{h}}{\partial \tilde{x}^2} \quad \text{at} \quad \tilde{y} = \tilde{h}(\tilde{x},\tilde{t}), \tag{1.62}$$

$$\frac{\partial \tilde{u}}{\partial \tilde{y}} = -\frac{d\tilde{T}^{\mathrm{i}}}{d\tilde{x}} \quad \text{at} \quad \tilde{y} = \tilde{h}(\tilde{x},\tilde{t}). \tag{1.63}$$

In the second equation we neglected the term $\gamma \Delta T \tilde{T}/\sigma_0$, which in dimensional terms corresponds to replacing $\sigma(T)$ with σ_0 in the normal stress balance.

If the Bond number is negligibly small, then based on (1.57) the pressure is not a function of $\tilde{y}$ and therefore (1.60) can be integrated twice to give the velocity profile

$$\tilde{u} = \frac{1}{2}\frac{\partial \tilde{p}}{\partial \tilde{x}}\left(\tilde{y}^2 - 2\tilde{y}\tilde{h}\right) - \frac{d\tilde{T}^{\mathrm{i}}}{d\tilde{x}}\tilde{y}. \tag{1.64}$$

Here we used (1.63) and the no-slip condition, $\tilde{u} = 0$ at $\tilde{y} = 0$, to find the constants of integration.

Integration of (1.58) in $\tilde{y}$ results in the following equation:

$$\int_0^{\tilde{h}} \frac{\partial \tilde{u}}{\partial \tilde{x}}d\tilde{y} + \tilde{v}\Big|_0^{\tilde{h}} = 0, \tag{1.65}$$

which can be combined with (1.61) using the Leibniz rule (for differentiation under the integral sign) and the condition $\tilde{v} = 0$ at $\tilde{y} = 0$, resulting in

$$\frac{\partial \tilde{h}}{\partial \tilde{t}} + \frac{\partial}{\partial \tilde{x}}\left(\int_0^{\tilde{h}} \tilde{u}d\tilde{y}\right) = 0. \tag{1.66}$$

This equation can also be derived by considering an integral mass balance for a film region between $\tilde{x}$ and $\tilde{x} + \Delta\tilde{x}$ in the limit of $\Delta\tilde{x} \to 0$.

Substituting the velocity profile (1.64) into (1.66) we obtain the following equation for film thickness $\tilde{h}$:

$$\frac{\partial \tilde{h}}{\partial \tilde{t}} + \frac{\partial}{\partial \tilde{x}}\left(\frac{\tilde{h}^3}{3}\frac{\partial^3 \tilde{h}}{\partial \tilde{x}^3} - \frac{\tilde{h}^2}{2}\frac{d\tilde{T}^{\mathrm{i}}}{d\tilde{x}}\right) = 0. \tag{1.67}$$

Here we used (1.62) to express $\tilde{p}$ in terms of the interface shape. The nondimensional interfacial temperature $\tilde{T}^{\mathrm{i}}$ can be expressed in terms of $\tilde{x}$ and the parameter $\tilde{\alpha} = \alpha d^2 \mathrm{Ca}^{-2/3}$ using (1.39), resulting in

$$\tilde{T}^{\mathrm{i}} = \mathrm{e}^{-\tilde{\alpha}\tilde{x}^2}. \tag{1.68}$$

Equation (1.67) is referred to as an *evolution equation* for film thickness since it describes evolution of the thickness in time and space provided that proper boundary and initial conditions are specified. Since the interface deformation is caused by temperature gradients and the derivative of the profile given by (1.68) is small for $\tilde{x} \gg \tilde{\alpha}^{-1/2}$, it is natural to expect the deformation to be small as well for sufficiently large $\tilde{x}$. We consider (1.67) on the domain $[0,L]$, $L = 20\tilde{\alpha}^{-1/2}$, so that the film is flat near $\tilde{x} = L$, justifying the boundary conditions

$$\tilde{h}(L,\tilde{t}) = 1, \qquad \frac{\partial \tilde{h}}{\partial \tilde{x}}(L,\tilde{t}) = 0. \tag{1.69}$$

Based on (1.68), we expect the solution to be an even function of $\tilde{x}$, so we consider only positive values of $\tilde{x}$ and impose two symmetry conditions at the left boundary of the computational domain,

$$\frac{\partial \tilde{h}}{\partial \tilde{x}}(0,\tilde{t}) = 0, \qquad \frac{\partial^3 \tilde{h}}{\partial \tilde{x}^3}(0,\tilde{t}) = 0. \tag{1.70}$$

Finally, since the film is initially uniform,

$$\tilde{h}(\tilde{x},0) = 1, \qquad 0 \le \tilde{x} \le L. \tag{1.71}$$

Equation (1.67) with the conditions (1.69)–(1.71) is solved numerically using the approach described below in Sect. 1.6. The numerical solution allows one to find the interface shape as a function of time. The result is usually represented by plots showing the snapshots of the interface at different moments in time. An example of such plot is shown in Fig. 1.7. By comparing different curves in the figure, we observe that the liquid film becomes thinner in the middle of the heated area as the thermocapillary flow pushes liquid away from there and into a ridge formed in a colder part of the film. Note that the results are plotted in scaled variables, so the actual interface deformation in the vertical direction is much smaller than the unit length in the horizontal direction.

It is clear in Fig. 1.7 that the region of significant interface deformation is initially localized near $\tilde{x} = 0$ and is gradually expanding. Since the extent of the film in our mathematical model is assumed to be infinite, the computational domain size L has to be sufficiently large so that the simulation results are independent of the choice of L. Indeed, the numerical simulations conducted for L between 20 and 25 show no significant differences in the values of $\tilde{h}(\tilde{x},\tilde{t})$ for the time interval considered ($0 \le \tilde{t} \le 10$). Plots of the interface shapes generated for these different computational domains (not shown) indicate that the interface remains nearly flat for $\tilde{x} > 12$. Since

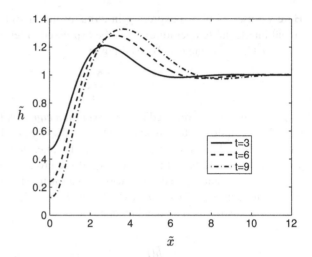

Fig. 1.7 Snapshots of the liquid–air interface obtained from the numerical solution of (1.67) with the interfacial temperature profile given by (1.68), $\tilde{\alpha} = 1$

the main purpose in Fig. 1.7 is to show the interface in the region where it does get deformed, the values of $\tilde{x}$ are restricted to the domain $[0, 12]$.

As the thermocapillary flow develops, the minimum film thickness gradually decreases until the liquid–air interface touches the solid surface, i.e., until the film ruptures. The evolution of the film at later times is no longer described by the model of the present section since the appearance and evolution of a dry area of the solid is not accounted for in our formulation. Mathematical models capable of describing these types of physical phenomena are introduced and discussed in detail in Chap. 2.

The Gaussian temperature profile used in our solution of (1.67) was chosen as a simple model of localized heating of the interface, but it turns out to be relevant for practical applications [7, 139]. For example, a pulsed laser with a Gaussian irradiance pattern focused to the surface of a uniform film results in a similar temperature distribution pattern. The film is initially solid (e.g., metal) but melts rapidly due to laser-induced heating and then flows away from the center of the laser spot due to thermocapillarity. This process can be used to create an axisymmetric hole which persists after the laser is turned-off and the film is resolidified. The solid substrate under the film, e.g., glass, does not melt in the process. If the laser beam is moving, a long channel can be fabricated in the film using the same principle.

To complete our discussion of the lubrication-type approach introduced in the present section, it is useful to mention a slightly different and somewhat more formal approach leading to the evolution equation (1.67). Instead of deriving the dimensional simplified equations (1.46)–(1.48), one can start by rewriting the full equations (1.41)–(1.43) with appropriate general boundary conditions in nondimensional form and then use asymptotic expansions in powers of ε (assuming $\mathrm{Ca} \sim \varepsilon^3$) for the scaled quantities such as pressure, velocity components, and film thickness, e.g.,

$$\tilde{p} = \tilde{p}_0 + \varepsilon \tilde{p}_1 + \varepsilon^2 \tilde{p}_2 + \cdots. \tag{1.72}$$

Then, using the terminology of the perturbation theory [71], our derivation of (1.67) corresponds to solving the order of ε problem. One can obtain higher-order corrections to our solution by considering the equations and boundary conditions at orders of ε^2, ε^3, etc., but such corrections are rarely considered in the models of thin liquid films.

1.5* Stokes Flow Equations and Interfacial Boundary Conditions

The lubrication-type equations (1.46)–(1.48) derived in the previous section are applicable only if the ratio ε of the length scales in the directions of the two coordinate axes is small. The purpose of the present section is to discuss modeling approaches for situations when this assumption is no longer valid. As an example, one can think of a liquid film on a solid substrate with the liquid surface temperature varying on the scale comparable to the film thickness, although the arguments used in the present section are applicable to more complicated flow configurations as well. For simplicity, we once again restrict our attention to two-dimensional flows, i.e., assume that the liquid velocity components, u and v, are functions of time t and the Cartesian coordinates x and y, but not z. As in the previous section, we start with the general system of dimensional equations (1.41)–(1.43). However, the length scales in the x- and y-directions are now assumed to be the same, denoted by L, and so are the velocity scales, denoted by U. The natural time scale under these conditions is L/U. In most microscale applications, the Reynolds number, $\mathrm{Re} = UL/\nu$, is small, so the following estimate holds,

$$\frac{|u_t + uu_x + vu_y|}{\nu|u_{xx} + u_{yy}|} \sim \frac{U^2/L}{\nu U/L^2} = \mathrm{Re} \ll 1. \tag{1.73}$$

The general system (1.41)–(1.43) can then be simplified to the so-called *Stokes flow equations*, widely used in the studies of microscale flows:

$$\frac{\partial p}{\partial x} = \mu \left(\frac{\partial^2 u}{\partial x^2} + \frac{\partial^2 u}{\partial y^2} \right) + f_x, \tag{1.74}$$

$$\frac{\partial p}{\partial y} = \mu \left(\frac{\partial^2 v}{\partial x^2} + \frac{\partial^2 v}{\partial y^2} \right) + f_y, \tag{1.75}$$

$$\frac{\partial u}{\partial x} + \frac{\partial v}{\partial y} = 0. \tag{1.76}$$

Here $f_x = 0$, $f_y = -\rho g$ for the liquid film on a horizontal substrate, but the equations are valid for arbitrary body force components, f_x and f_y.

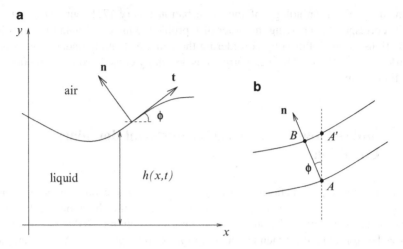

Fig. 1.8 (**a**) Sketch of a curved liquid–air interface showing the Cartesian coordinates, the unit normal and tangential vectors to the interface, and the slope angle. (**b**) Sketch for the derivation of the kinematic boundary condition

Let us now discuss the boundary conditions for the Stokes flow equations at a curved liquid–air interface, shown in Fig. 1.8a, with $h(x,t)$ denoting the distance from the interface to the x-axis. The function $h(x,t)$ can represent the entire interface, e.g., for the case of a liquid film on a flat solid substrate, or part of the interface for more complicated geometric configurations. Since the tangent of the slope angle ϕ is equal to h_x, the unit normal and tangential vectors **n** and **t**, shown in Fig. 1.8a, can be expressed in terms of the interface shape as:

$$\mathbf{n} = (-h_x, 1)\left(1 + h_x^2\right)^{-1/2}, \qquad \mathbf{t} = (1, h_x)\left(1 + h_x^2\right)^{-1/2}. \tag{1.77}$$

Figure 1.8(b) shows the snapshots of an interface segment at two different moments in time separated by a small interval Δt. The unit normal vector **n** at an arbitrary point A on the lower curve intersects with the upper curve (corresponding to interface location at the later time) at a point B and the vertical dashed line through the point A intersects with the upper curve at a point A'. The angle ϕ shown in the sketch in Fig. 1.8b is equal to the local slope angle of the interface. In the limit of $\Delta t \to 0$,

$$\cos\phi = \frac{|AB|}{|AA'|}, \qquad |AB| = V_n\Delta t, \qquad |AA'| = h_t\Delta t, \tag{1.78}$$

where V_n is the normal velocity of the interface motion. Using $\tan\phi = h_x$ and (1.78), we obtain a formula for the normal velocity,

$$V_{\mathrm{n}} = \frac{h_t}{\sqrt{1 + h_x^2}}. \tag{1.79}$$

At the interface, the flow velocity in the direction of $\mathbf{n}$ is the same as V_{n}, resulting in the kinematic boundary condition,

$$h_t + u h_x - v = 0. \tag{1.80}$$

Forces acting in the liquid are described using the stress tensor [1, 12]. In our two-dimensional model, only four components of this tensor enter the equations. These components can be written in a fixed coordinate system as:

$$\mathbf{T} = \begin{pmatrix} -p + 2\mu u_x & \mu(u_y + v_x) \\ \mu(u_y + v_x) & -p + 2\mu v_y \end{pmatrix}, \tag{1.81}$$

where p is the pressure in the liquid.

Using the same control volume as in Sect. 1.1 (shown in Fig. 1.1b), we once again observe that the total vector sum of all forces acting on the control volume should be zero. Considering the projections onto the normal and tangential directions, we obtain two conditions for the force components: the normal stress balance,

$$\mathbf{n} \cdot \mathbf{T} \cdot \mathbf{n} + p_0 = \sigma \kappa, \tag{1.82}$$

and the tangential stress balance,

$$\mathbf{t} \cdot \mathbf{T} \cdot \mathbf{n} = \frac{\partial \sigma}{\partial s}. \tag{1.83}$$

Here air is assumed to be at a uniform pressure p_0 and the viscous stresses in the air are neglected. The expression $\mathbf{T} \cdot \mathbf{n}$ denotes a tensor–vector dot product, which in a fixed coordinate system can be calculated by the standard rules of matrix–vector multiplication and has the physical meaning of the stress vector in the direction of $\mathbf{n}$; $\partial / \partial s$ denotes the derivative along the interface (s being the arclength variable), and κ is the curvature.

The stress balances can be written in terms of the interface shape using (1.77), (1.81), and the standard formula for curvature, resulting in

$$p_0 - p + 2\mu \frac{u_x h_x^2 + v_y - h_x(u_y + v_x)}{1 + h_x^2} = \sigma \frac{h_{xx}}{(1 + h_x^2)^{3/2}}, \tag{1.84}$$

$$\mu \frac{2h_x(v_y - u_x) + (u_y + v_x)(1 - h_x^2)}{1 + h_x^2} = \frac{\partial \sigma}{\partial s}. \tag{1.85}$$

Equations (1.80), (1.84), and (1.85) complete our formulation of the general set of boundary conditions at a deforming two-dimensional liquid–air interface. For the case of a thin liquid film on a flat substrate discussed in Sect. 1.4.2, the stress conditions can be simplified as follows. If the characteristic film thickness is d and the length scale in the x-direction, d/ε, is much larger than d, then the right-hand-side of (1.84) is of the order of $\varepsilon^2\sigma/d$. Following the same argument as in Sect. 1.4.2, if the characteristic value of u is denoted by U, the other velocity component, v, is on the order of εU and therefore

$$\mu\left|\frac{u_x h_x^2 + v_y - h_x(u_y + v_x)}{1 + h_x^2}\right| \sim \frac{\varepsilon\mu U}{d}. \tag{1.86}$$

These terms are negligible compared to the right-hand-side of (1.84) as long as Ca $\ll \varepsilon$. Thus, the standard assumptions of the lubrication-type model of liquid film, $\varepsilon \ll 1$ and Ca $\sim \varepsilon^3$, allow one to simplify (1.84) to

$$p - p_0 = -\sigma h_{xx}. \tag{1.87}$$

Note that the curvature is approximated here by the second derivative of h since for small ε the denominator of the right-hand-side of (1.84) can be replaced by unity. In the limit of small ε the left-hand-side of (1.85) can be replaced with μu_y since all other terms are much smaller. Then, (1.85) takes the form

$$\mu u_y = \frac{\partial\sigma}{\partial s}. \tag{1.88}$$

In addition to the boundary conditions at the fluid interface, our flow model in Sect. 1.4.2 included the conditions at the solid surface. If the Cartesian coordinates are such that the x-axis is along the solid and the y-axis is normal to it, then the velocity components satisfy

$$u = 0, \quad \text{at} \quad y = 0, \tag{1.89}$$

$$v = 0, \quad \text{at} \quad y = 0. \tag{1.90}$$

While the second of these conditions simply states that the liquid does not penetrate into the solid, the first one, called the no-slip condition, is actually an assumption. Even though it has been verified for a number of liquid–solid combinations by careful experimental studies reviewed, e.g., by Lauga et al. [85], as well as by more recent ones [90], some experimental data can only be described by a more general *Navier slip condition*, which in our Cartesian coordinates (with the y-axis pointing into the liquid) is written as:

$$u = \lambda\frac{\partial u}{\partial y} \qquad \text{at} \quad y = 0. \tag{1.91}$$

Here u is the (dimensional) horizontal velocity component in the x-direction, and λ is a constant called the slip length. The latter can be interpreted as the distance

Fig. 1.9 Sketch illustrating
the concept of slip length λ in
the Navier slip condition,
(1.91)

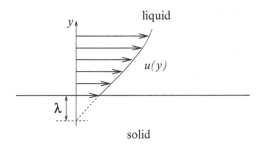

below the solid–liquid interface at which the velocity $u(y)$ extrapolates to zero, as
sketched in Fig. 1.9. For example, Cottin-Bizonne et al. [38] report the value of slip
length of 19 ± 2 nm in their study of flow of water near a Pyrex surface coated with
OTS (an organometallic chemical compound). Note that the static contact angle for
water on this surface is above $90°$, meaning that the surface is hydrophobic. Slip
lengths of tens of nanometers have been measured for flow of water near several
other hydrophobic surfaces. The physical mechanisms of slip for smooth solid
surfaces are still not completely understood, with possible explanations including
formation of a layer of gas or small-scale bubbles between the liquid and the solid
[85, 132]. Significant slip lengths can be achieved for flows near structured surfaces,
as discussed in detail below in Chap. 5.

It is interesting to note that liquid transport rates in microscale devices can be
increased due to slip effects. As an illustration, let us revisit the model of pressure-
driven flow in a channel of height $2d$ from Sect. 1.3. Solving (1.28) with the Navier
slip conditions at the walls of the channel (i.e., at $y = \pm d$) leads to the velocity
profile in the form

$$u = \frac{G}{2\mu}(d^2 - y^2) + \frac{Gd\lambda}{\mu} \tag{1.92}$$

and the corresponding average flow velocity

$$\bar{u} = \frac{Gd^2}{3\mu}\left(1 + 3\frac{\lambda}{d}\right). \tag{1.93}$$

For a typical microfluidic device ($d = 10\,\mu m$) and $\lambda = 20$ nm, the average velocity
increase compared to the solution with the no-slip condition at the walls is only
0.6%, but the effect can be more significant for smaller scale devices or for solid
walls with unusually large slip lengths.

Finally, let us briefly discuss the more general case of a two-dimensional
interface Γ between two viscous fluids in a three-dimensional setting. To make the
formulas more compact, we use x_i ($i = 1, 2, 3$) for the Cartesian coordinates and $u_i^{(k)}$
($i = 1, 2, 3$) for the corresponding velocity components of fluid 1 ($k = 1$) and fluid

Fig. 1.10 Sketch of the curve
C and the control volume δV
used in the derivation of the
stress conditions at a
two-dimensional fluid
interface Γ. Fluid 1 is below
the interface Γ and fluid 2 is
above it

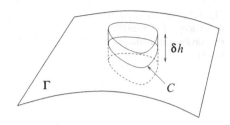

2 ($k = 2$). Both normal and tangential velocities of the two fluids are continuous at
the interface, meaning that

$$u_i^{(1)} = u_i^{(2)}, \quad \mathbf{x} \in \Gamma. \tag{1.94}$$

The stress tensors and the rate-of-strain tensors [1] are expressed in index
notation as:

$$T_{ij}^{(k)} = -p^{(k)}\delta_{ij} + 2\mu^{(k)}e_{ij}^{(k)}, \quad e_{ij}^{(k)} = \frac{1}{2}\left(\frac{\partial u_i^{(k)}}{\partial x_j} + \frac{\partial u_j^{(k)}}{\partial x_i}\right), \tag{1.95}$$

where $\delta_{ij} = 1$ for $i = j$ and zero otherwise, according to the standard convention.

To describe the effects of surface tension, consider an arbitrary closed curve C
enclosing a portion of the fluid interface Γ as sketched in Fig. 1.10. The total force
due to surface tension acting on this portion of the interface at its boundary is given
by the following line integral,

$$F_i^s = \int_C T_{ij}^s t_j \, \mathrm{d}l. \tag{1.96}$$

Here t_j denotes the components of a unit tangent vector to the interface Γ in the
direction perpendicular to the curve C, summation over repeated indices is implied,
and the tensor T_{ij}^s is defined in terms of the surface tension σ and the components
of the unit normal vector $\mathbf{n}$ as:

$$T_{ij}^s = \sigma(\delta_{ij} - n_i n_j). \tag{1.97}$$

The vector $\mathbf{n}$ is pointing into fluid 2. To verify that (1.97) can be used to obtain the
correct expression for the surface tension force we observe that

$$T_{ij}^s n_j = 0, \quad T_{ij}^s t_j = \sigma t_i \tag{1.98}$$

since $n_j n_j = 1$, $n_j t_j = 0$. While the surface tension σ and the normal vector $\mathbf{n}$ in the
above formulas are defined only at the interface Γ, it is convenient to redefine them
as functions of all three spatial variables everywhere in the vicinity of Γ using the

following procedure. Consider an arbitrary point P at a distance Δ from the interface Γ and suppose a point P' at the interface is the closest one to P. For a smooth surface Γ in the limit of small Δ, there is only one such point P' for every point P. The value of the function $\sigma(\mathbf{x})$ at P is then defined as equal to the local surface tension at the point P'. This definition implies that the function $\sigma(\mathbf{x})$ does not vary in the direction normal to the interface. The same approach is used to redefine the normal vector $\mathbf{n}$ as a function of all three spatial coordinates near Γ. Thus, the functions $\sigma(\mathbf{x})$, $\mathbf{n}(\mathbf{x})$ and, using (1.97), the tensor T_{ij}^s are now defined everywhere in a thin layer near the interface.

For an arbitrary point on the curve C, consider a line segment of length δh centered at that point and normal to the interface Γ; the set of all such line segments makes up the side boundary of the control volume δV shown in the sketch in Fig. 1.10. The top and bottom boundaries of δV are at a distance $\delta h/2$ above and below the interface Γ, respectively. Consider an integral of the quantity $T_{ij}^s \hat{n}_j$ over the entire boundary of the control volume δV, where $\hat{\mathbf{n}}$ is the outward unit normal vector at the boundary. At the top and bottom bounding surfaces of δV, the outward normal $\hat{\mathbf{n}}$ is equal to $\mathbf{n}$ and $-\mathbf{n}$, respectively, so $T_{ij}^s \hat{n}_j = 0$ there. Since the quantity $T_{ij}^s \hat{n}_j$ does not vary in the direction normal to the interface Γ, the integral of this quantity over the side boundary of the control volume in the limit of small δh can be replaced with the line integral along C multiplied by δh. Note that the outward normal vector $\hat{\mathbf{n}}$ along C is the same as the previously defined unit tangent vector with the components t_j. Application of the divergence theorem then leads to

$$\delta h \int_C T_{ij}^s t_j dl = \int_{\delta V} \frac{\partial T_{ij}^s}{\partial x_j} dV. \tag{1.99}$$

Observing that the divergence of T_{ij}^s does not vary in the direction normal to the interface, the volume integral on the right-hand-side of (1.99) can be replaced with an integral over the surface S (the portion of Γ enclosed by C) multiplied by δh, leading to the surface analog of the classical divergence theorem,

$$\int_C T_{ij}^s t_j dl = \int_S \frac{\partial T_{ij}^s}{\partial x_j} dS. \tag{1.100}$$

In the limit of $\delta h \to 0$, the mass and momentum of the liquid inside the control volume δV (shown in Fig. 1.10) approach zero, so the vector sum of all forces acting on the control volume should be zero. The contributions to the force balance which remain finite in the limit of infinitely small δh are the force $\mathbf{F}^s$, due to the surface tension acting on the side boundary of the control volume, and the forces in the two fluids acting at the top and bottom boundaries, so

$$F_i^s = \int_S \left(T_{ij}^{(1)} - T_{ij}^{(2)} \right) n_j dS. \tag{1.101}$$

Fig. 1.11 Sketch of a
two-dimensional interface Γ
showing the local Cartesian
coordinates used to express
the normal stress condition in
terms of mean curvature. The
dashed line is the curve of
intersection of the interface Γ
and the ξ_1–ξ_3 plane

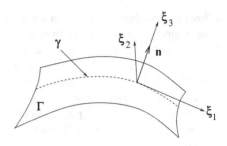

Using (1.96) and (1.100) to express the force $\mathbf{F}^s$ as an integral over S and recalling
that the region S is arbitrary, we obtain the following boundary condition at the fluid
interface

$$\left(T_{ij}^{(1)} - T_{ij}^{(2)} \right) n_j = \frac{\partial T_{ij}^s}{\partial x_j}, \qquad \mathbf{x} \in \Gamma. \tag{1.102}$$

Using the definition of the tensor T_{ij}^s given by (1.97) and recalling that $\sigma(\mathbf{x})$ and
$\mathbf{n}(\mathbf{x})$ do not vary in the direction normal to the interface, the interfacial boundary
condition is transformed to

$$\left(T_{ij}^{(1)} - T_{ij}^{(2)} \right) n_j = \frac{\partial \sigma}{\partial x_i} - \sigma \frac{\partial n_j}{\partial x_j} n_i, \qquad \mathbf{x} \in \Gamma. \tag{1.103}$$

By considering projections of this equation onto the normal and a tangential
direction defined by an arbitrary tangent vector, $\mathbf{t}$, we obtain the normal and shear
stress balances at the interface in the form

$$n_i \left(T_{ij}^{(1)} - T_{ij}^{(2)} \right) n_j = -\sigma \frac{\partial n_i}{\partial x_i}, \qquad \mathbf{x} \in \Gamma, \tag{1.104}$$

and

$$t_i \left(T_{ij}^{(1)} - T_{ij}^{(2)} \right) n_j = t_i \frac{\partial \sigma}{\partial x_i}, \qquad \mathbf{x} \in \Gamma. \tag{1.105}$$

To understand the geometric meaning of the divergence of the normal vector on
the right-hand-side of (1.104), consider an arbitrary point on the interface and local
Cartesian coordinates ξ_i with the origin at this point and such that ξ_3 is along the
local normal vector to the interface, as shown in the sketch in Fig. 1.11. The dashed
line γ in the sketch, defined as the intersection of the ξ_1–ξ_3 plane and the interface
Γ, has curvature κ at the origin of the local coordinate system. Note that γ is a planar
curve so its curvature is defined as the derivative of the slope angle ϕ with respect
to the arclength s. Since the ξ_1-axis is in the direction of a local tangent to the curve
γ, we can replace ds with dξ_1 so that

$$\kappa = \frac{d\phi}{d\xi_1}. \tag{1.106}$$

If the coordinate system is rotated around the ξ_3-axis, the curvature κ will change as a function of the rotation angle in a continuous fashion, so there is a particular choice of the rotation angle such that the curvature κ reaches its minimum value, denoted by κ_1. From now on, the orientation of the local coordinate system is assumed to be such that ξ_1 is in the direction corresponding to the minimum curvature κ_1.

For an arbitrary point on the curve γ, consider the unit normal vector to the two-dimensional interface Γ, with components in the local coordinate system denoted by n'_i and the projection onto the ξ_1–ξ_3 plane denoted by $\mathbf{n}^{\|}$. The latter is also a normal vector to the curve γ, so its components can be expressed in terms of the local slope angle of γ as:

$$n'_1 = -|\mathbf{n}^{\|}|\sin\phi, \qquad n'_3 = |\mathbf{n}^{\|}|\cos\phi. \tag{1.107}$$

Differentiating the first of these formulas with respect to ξ_1 and then using the second one together with the conditions of $n'_3 = 1$ and $\phi = 0$ at the origin leads to

$$\frac{\partial n'_1}{\partial \xi_1} = -\kappa_1. \tag{1.108}$$

Here we also used (1.106) to express the final result in terms of the local curvature of γ at the origin, equal to κ_1 due to the choice of the orientation of the local coordinate system.

Based on a well-known result from differential geometry (see, e.g., Chap. 5 in [96]), if the direction of ξ_1 corresponds to the minimum value κ_1 of the curvature, then the maximum value κ_2 is reached in the orthogonal direction, i.e., in the direction of ξ_2. The quantities κ_1 and κ_2 are referred to as the principal curvatures while their average,

$$H = \frac{\kappa_1 + \kappa_2}{2}, \tag{1.109}$$

is called the mean curvature. Using the same arguments as in the derivation of (1.108) we find that

$$\frac{\partial n'_2}{\partial \xi_2} = -\kappa_2. \tag{1.110}$$

Combining (1.108), (1.110), and the definition of the mean curvature, we obtain

$$\frac{\partial n'_1}{\partial \xi_1} + \frac{\partial n'_2}{\partial \xi_2} = -2H. \tag{1.111}$$

Here the left-hand-side of the equation represents the divergence of $\mathbf{n}$ (there is no variation of $\mathbf{n}$ in the ξ_3-direction). Since the divergence of a vector field is independent from the choice of the coordinate system, we immediately obtain $\frac{\partial n_i}{\partial x_i} = -2H$, so (1.104) takes the form

$$n_i \left(T_{ij}^{(1)} - T_{ij}^{(2)} \right) n_j = 2\sigma H, \qquad \mathbf{x} \in \Gamma. \tag{1.112}$$

For a static configuration, this formula reduces to the classical Young–Laplace equation,

$$p^{(1)} - p^{(2)} = -2\sigma H, \quad \mathbf{x} \in \Gamma. \tag{1.113}$$

To justify that $\partial n_3' / \partial \xi_3 = 0$ in the above derivation, we noted that the vector field $\mathbf{n}$ does not vary in the direction normal to the interface. Later in the present section, we use a vector field $\bar{\mathbf{n}}$ which is identical to $\mathbf{n}$ at the interface Γ but can vary in the direction normal to Γ. It turns out that for any such smoothly varying vector field with $|\bar{\mathbf{n}}| = 1$, the condition of $\partial \bar{n}_3' / \partial \xi_3 = 0$ is satisfied, where $\bar{n}_3'$ denotes the component of $\bar{\mathbf{n}}$ in the ξ_3-direction. Indeed, since $\bar{n}_3'$ is equal to unity at $\xi_3 = 0$, the maximum of $\bar{n}_3'$ along the ξ_3-direction is reached at that point and therefore $\partial \bar{n}_3' / \partial \xi_3 = 0$ at $\xi_3 = 0$. The condition of $\mathbf{n} = \bar{\mathbf{n}}$ everywhere along the interface Γ immediately leads to

$$\frac{\partial n_1'}{\partial \xi_1} = \frac{\partial \bar{n}_1'}{\partial \xi_1}, \quad \frac{\partial n_2'}{\partial \xi_2} = \frac{\partial \bar{n}_2'}{\partial \xi_2}. \tag{1.114}$$

Thus, the divergence of $\mathbf{n}$ can be replaced with the divergence of $\bar{\mathbf{n}}$ when calculating the mean curvature.

While the index notation is convenient for deriving the general interfacial boundary conditions, in practical applications the Cartesian coordinates are usually denoted by x, y, and z and two-dimensional interface shapes are represented by functions of the form $z = h(x,y,t)$, where t is the time variable. For an arbitrary point (x_0, y_0, z_0) at the interface at a given time, the equation of the local tangent plane is

$$z - z_0 = h_x(x_0, y_0, t)(x - x_0) + h_y(x_0, y_0, t)(y - y_0). \tag{1.115}$$

Based on this equation, the local unit normal vector to the interface can be expressed in terms of its shape as:

$$\bar{\mathbf{n}} = \frac{(-h_x, -h_y, 1)}{\sqrt{1 + h_x^2 + h_y^2}}. \tag{1.116}$$

The mean curvature is then given by:

$$H = -\frac{1}{2} \nabla \cdot \bar{\mathbf{n}}, \tag{1.117}$$

where $\nabla = \left(\frac{\partial}{\partial x}, \frac{\partial}{\partial y}, \frac{\partial}{\partial z} \right)$. Assuming that (1.116) is valid not only at the interface but also in its vicinity, the divergence of the vector field $\bar{\mathbf{n}}$ in (1.117) can be calculated, leading to the following expression for the mean curvature,

$$H = \frac{(1 + h_y^2)h_{xx} - 2h_x h_y h_{xy} + (1 + h_x^2)h_{yy}}{2(1 + h_x^2 + h_y^2)^{3/2}}. \tag{1.118}$$

For an important special case of a surface of revolution, described by a function $z = h(r,t)$ in cylindrical coordinates, the mean curvature is given by:

$$H = \frac{h_{rr}}{2(1 + h_r^2)^{3/2}} + \frac{h_r}{2r(1 + h_r^2)^{1/2}}. \tag{1.119}$$

The two terms on the right-hand-side of this equation represent the two principal curvatures. The first of them is simply the curvature of the planar curve $h(r,t)$ while the second one is derived using a simple geometric construction discussed, e.g., in Sect. II.2 of Adamson & Gast [2].

1.6 Numerical Solution of Evolution Equations

The key advantage of the lubrication-type approach formulated in Sect. 1.4 is that the problem of describing the liquid flow is reduced to solving a single nonlinear evolution equation for the liquid film thickness, such as, (1.67). However, evolution equations typically do not allow for simple analytical solutions and therefore have to be solved numerically. Let us briefly discuss a simple numerical method, often called the method of lines, which allows one to obtain interface shapes such as the ones illustrated in Fig. 1.7. The idea of the method is to reduce the problem to solving a system of ordinary differential equations for time-dependent values of the film thickness at a discrete set of values of the spatial coordinate $\tilde{x}$. While this approach is applicable to a wide range of problems, we illustrate it here by considering (1.67) on the domain $[0, L]$ with the boundary and initial conditions given by (1.69)–(1.71). It is convenient to rewrite (1.67) as:

$$\frac{\partial \tilde{h}}{\partial \tilde{t}} = -\frac{\partial Q}{\partial \tilde{x}}, \qquad Q = \frac{\tilde{h}^3}{3} \frac{\partial^3 \tilde{h}}{\partial \tilde{x}^3} - \frac{\tilde{h}^2}{2} \frac{d\tilde{T}^i}{d\tilde{x}}. \tag{1.120}$$

We discretize the domain by dividing it into N equal sub-intervals, each of length $\Delta \tilde{x} = L/N$, as sketched in Fig. 1.12, so that the coordinates of the grid points are $\tilde{x}_i = (i-1)\Delta \tilde{x}$, $i = 1, \ldots, N$. Film thickness values at the grid points, denoted by $\tilde{h}_i$, are functions of time only and satisfy a system of N first-order ordinary differential equations,

$$\frac{\partial \tilde{h}_i}{\partial \tilde{t}} = -\frac{Q_{i+1/2} - Q_{i-1/2}}{\Delta \tilde{x}}, \tag{1.121}$$

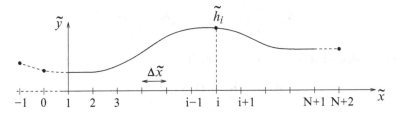

Fig. 1.12 Sketch of the discretization used in the numerical solution of (1.67)

where $Q_{i\pm1/2}$ are the approximate values of Q at the points half-way between the grid points, $\tilde{x}_{i\pm1/2} = \tilde{x}_i \pm \Delta\tilde{x}/2$. For the values of i from 3 to $N-2$ we use

$$Q_{i+1/2} = \frac{\tilde{h}_i^3 + \tilde{h}_{i+1}^3}{6(\Delta\tilde{x})^3} \left(\tilde{h}_{i+2} - 3\tilde{h}_{i+1} + 3\tilde{h}_i - \tilde{h}_{i-1} \right) + \frac{\tilde{h}_i^2 + \tilde{h}_{i+1}^2}{2(\Delta\tilde{x})^2} \tilde{\alpha}\tilde{x}_{i+1/2} e^{-\tilde{\alpha}\tilde{x}_{i+1/2}^2}$$

$$(1.122)$$

and a similar expression for $Q_{i-1/2}$. In the first term on the right-hand-side of (1.122), the standard finite-difference approximation of the third derivative [130] at a point $\tilde{x}_{i+1/2}$ is used; the accuracy of this approximation is discussed in more detail below. Since $i = N+1$ corresponds to $\tilde{x} = L$, the value of the thickness at this endpoint according to (1.69) is equal to unity, so $\tilde{h}_{N+1} = 1$. Near the boundaries, one needs points outside the domain (the so-called ghost points, shown by filled circles in Fig. 1.12) to be able to evaluate the derivatives when using centered finite-differences. The values of the function at the ghost points are expressed in terms of the function values in the computational domain using the boundary conditions as follows. Since the film is expected to be flat near $\tilde{x} = L$, its thickness at a ghost point corresponding to $i = N+2$ can be approximated using the condition of zero slope, resulting in $\tilde{h}_{N+2} = \tilde{h}_N$. Near $\tilde{x} = 0$, we use the symmetry conditions (1.70) to obtain $\tilde{h}_0 = \tilde{h}_2$, $\tilde{h}_{-1} = \tilde{h}_3$. The system of equations (1.121) can then be solved using standard differential equation solvers. A MATLAB code for this problem can be found in Sect. B.1 in Appendix B. Typically, a wide range of different time scales is present in the numerical solutions of systems of ordinary differential equations obtained by discretizing evolution equations, meaning that these systems are stiff. It is therefore important to choose a time-stepping method capable of handling stiff systems. The code in Sect. B.1 uses the standard MATLAB solver ode15s for stiff systems of ordinary differential equations and produces Fig. 1.7 as its output.

It is important to emphasize that the function Q in (1.121) has to be evaluated at $i \pm 1/2$ as opposed to $i \pm 1$ to make sure that the discretization is consistent with the condition of conservation of mass of the liquid, which in our two-dimensional framework implies conservation of the area under the curve,

$$\int_0^L \tilde{h}\, d\tilde{x}.$$

$$(1.123)$$

This integral can be approximated numerically by the sum

$$S_N = \left(\frac{1}{2}\tilde{h}_1 + \sum_{i=2}^N \tilde{h}_i + \frac{1}{2}\tilde{h}_{N+1} \right) \Delta\tilde{x}.$$

$$(1.124)$$

By direct calculation using (1.121) we obtain

$$\frac{dS_N}{d\tilde{t}} = \frac{Q_{3/2} + Q_{1/2}}{2} - Q_{N+1/2}.$$

$$(1.125)$$

The first term on the right-hand-side of this equation is equal to zero due to the symmetry conditions at $\tilde{x} = 0$ and the second term is negligibly small since $i = N + 1/2$ corresponds to the region where the film is flat and the liquid is at rest. Thus, the value of S_N is conserved by our numerical method.

The right-hand-side of (1.121) represents an approximation of the actual quantity $-\partial Q/\partial \tilde{x}$. It is natural to ask how accurate this approximation is and, even more importantly, how quickly the accuracy improves as the number of grid points is increased. The values of $\tilde{h}$ at the grid points around i can be expressed in terms of the (unknown) exact values of derivatives at $\tilde{x}_i = (i-1)\Delta\tilde{x}$ by writing, e.g.,

$$\tilde{h}_{i+1} = \sum_{n=0}^{\infty} \frac{\tilde{h}^{(n)}(\tilde{x}_i)}{n!}(\Delta\tilde{x})^n. \tag{1.126}$$

Direct substitution of expressions such as this one into (1.122) and a similar formula for $Q_{i-1/2}$ leads to expressions which are then used to estimate the right-hand side of (1.121). The result shows that for small $\Delta\tilde{x}$ the largest term in the difference between the exact expression for $-\partial Q/\partial \tilde{x}$ and its approximation is proportional to $(\Delta\tilde{x})^2$. Thus, our finite-difference discretization is *second-order accurate* in space. To illustrate that this prediction is indeed consistent with the numerical results we run our code for several different values of N and evaluate the norm of the error for the numerical solution of the discretized system,

$$E = \left(\Delta\tilde{x}\sum_{i=1}^{N}\left(\tilde{h}_i^N - \tilde{h}_i^f\right)^2\right)^{1/2}, \tag{1.127}$$

at a fixed time $t = 10$. Here $\tilde{h}_i^N$ and $\tilde{h}_i^f$ are the numerical solutions on the grid of N points and on the finest grid used in the simulation, respectively. The difference between $\tilde{h}_i^f$ and the (unknown) exact solution is assumed negligible when estimating the error norm. The tolerances in the time-stepping routine are chosen small enough so that the time-discretization part of the error is negligible compared to the contribution due to the approximation of spatial derivatives. The plot of the error norm as a function of $\Delta\tilde{x}$ in log–log coordinates in Fig. 1.13 shows that points corresponding to different grid sizes lie on a straight line. The slope of this line is equal to 2, meaning that the error norm decays as $(\Delta\tilde{x})^2$ in accordance with our theoretical prediction. Comparison between the numerical results for the error and its expected theoretical behavior is a useful tool for validation of numerical codes.

Numerical solution of evolution equations is easy to implement not only in MATLAB but also in Fortran or C/C++ since several efficient solvers for systems of ordinary differential equations are available online. In particular, the solvers VODE/DVODE and DASSL/DDASSL have been widely used to solve evolution equations [24, 99]. Spectral methods are often used instead of the finite-difference discretization of derivatives, especially for evolution equations with periodic boundary conditions, as discussed, e.g., in Canuto et al. [29]. The accuracy of the standard numerical methods for evolution equations can decrease dramatically when the

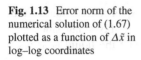

Fig. 1.13 Error norm of the
numerical solution of (1.67)
plotted as a function of $\Delta \tilde{x}$ in
log–log coordinates

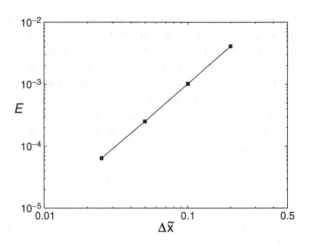

thickness of the liquid film at some grid points becomes very small compared to
its initial value. A numerical method which is free from this limitation has been
developed recently by Zhornitskaya and Bertozzi [148].

1.7   Disjoining Pressure in Ultra-Thin Liquid Films

Consider a gas bubble floating in a liquid below a smooth horizontal solid
surface. Due to buoyancy, the bubble will move upward until it reaches the solid
surface and an equilibrium configuration is established. Two qualitatively different
equilibrium configurations have been observed experimentally. In the first one, the
liquid–gas interface is in contact with the solid so that a circular dry patch is formed
on the solid surface. In this case, the static shape of the bubble can be found using
the approach described in Sect. 1.2 (except that the geometric configuration here
is axisymmetric rather than two-dimensional, so the equations of capillary statics
have to be formulated in cylindrical coordinates). In the second configuration the
gas phase and the solid remain separated by a very thin liquid film as shown in
Fig. 1.14a. Detecting such a film (and thus distinguishing the two configurations)
experimentally is challenging since the film thickness is typically below 1 μm and
can be as small as several nanometers. One approach to experimental investigation
of such ultra-thin films involves studying reflections of light from the solid–liquid
and liquid–gas interfaces for the case when the liquid–gas interface is partially
reflecting, as sketched in Fig. 1.14b. Two optical beams shown by the solid and
dashed lines in the sketch travel different distances, resulting in a phase shift
between them. By investigating this phase shift the value of film thickness can be
recovered. Changes in polarization of light during reflections at interfaces can also
be used to measure film thickness by a technique called ellipsometry. A much more
detailed discussion of experimental methods for studies of ultra-thin films can be
found in several books [45, 72, 122].

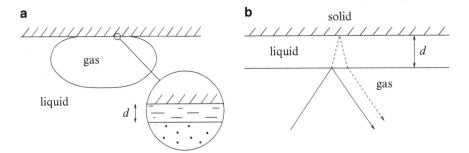

Fig. 1.14 (**a**) Sketch of a stationary gas bubble near a solid surface; (**b**) Illustration of the basic idea behind optical methods for measuring thickness of ultra-thin liquid films

The existence of a stable flat film in equilibrium with a curved meniscus, as sketched in Fig. 1.14a, seems to contradict the equation of capillary statics discussed in Sect. 1.2. Indeed, liquid pressure in the flat thin film, p, should be equal to the pressure in the gas inside the bubble, p_g, assumed uniform. Since there is no fluid flow, the pressure in the liquid near the meniscus and away from the solid is expected to be the same as p (in fact slightly higher if the effect of gravity is nonnegligible). However, the meniscus is curved, so the equation of capillary statics discussed in Sect. 1.2 requires the pressure in the liquid to be *lower* than $p_g = p$. This contradiction was resolved in the pioneering works of Derjaguin et al. [44,45]. They pointed out that unbalanced intermolecular interactions in very thin films can result in additional (compared to the bulk liquid) terms in the stress tensor and introduced the term "disjoining pressure" to describe these terms. When the disjoining pressure Π is taken into account, the normal stress in the film, $p + \Pi$, is equal to the pressure in the gas, p_g. If gravity is negligible (due to small values of the Bond number as discussed in Sect. 1.2), the pressure in the bulk liquid p_0 is equal to p and the disjoining pressure can be found from

$$\Pi = p_g - p_0. \tag{1.128}$$

A sketch in Fig. 1.15 illustrates how intermolecular interactions in ultra-thin films lead to disjoining pressure. Suppose a molecule marked "A" in the sketch has nonnegligible interaction with molecules inside a sphere of radius R, shown by a large circle in the cross-section sketch in the figure. In the bulk, only molecules of the same type as "A" are inside this sphere (assuming the liquid is pure). Averaging such interactions results in the standard macroscopic equations of fluid mechanics. However, the situation shown in Fig. 1.15 is different: the molecule "A" interacts with solid molecules (gray circles) and gas molecules (empty circles). As a result, intermolecular interaction energies are different and the standard macroscopic equations have to be modified. Specific formulas for calculating Π depend on the details of intermolecular interaction. If the electric charge and the average dipole moment of each molecule are zero, then the leading term in the potential

Fig. 1.15 Sketch illustrating the molecular basis of understanding the disjoining pressure

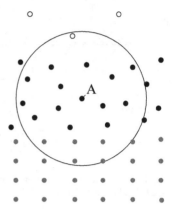

of interaction between molecules is due to fluctuations of dipole moment of each molecule and is proportional to inverse sixth power of the distance between them. The simplest method for estimating the total interaction energy is to sum up the pairwise interactions between molecules. However, the weaknesses of this approach are that it requires an unknown cut-off length to avoid diverging integrals and that it does not take into account the fact that interaction energy of several molecules is not a simple sum of their pairwise interactions. A theory free from these limitations, developed by Dzyaloshinskii et al. [47], treats liquid in the film as a continuous dielectric medium with fluctuating electromagnetic field. This approach allows one to express the disjoining pressure in terms of the dielectric permittivities of the liquid and solid. These quantities depend on the frequency of the electromagnetic field and have been investigated experimentally over a wide range of frequencies. The actual formulas obtained in Dzyaloshinskii et al. [47] are rather complicated, but they can be simplified for two important special cases. If the film thickness d is small compared to the characteristic wavelength λ in the spectrum of the electromagnetic field, then the disjoining pressure Π is inversely proportional to the cube of the thickness,

$$\Pi = -\frac{A}{d^3},\qquad(1.129)$$

where A is referred to here and below as the Hamaker constant (although some studies use the term "Hamaker constant" for the quantity $A' = 6\pi A$). Remarkably, the same dependence of disjoining pressure on the film thickness is obtained by pairwise summation of intermolecular interactions discussed above. Equation (1.129) is often referred to as a London–van der Waals formula or simply as a van der Waals model of disjoining pressure. In the opposite limit of $d \gg \lambda$, the effects of finite propagation speed of the electromagnetic waves (retardation effects) become important and the disjoining pressure is proportional to d^{-4}. The formulas for disjoining pressure applicable at intermediate values of thickness $d \sim \lambda$ are discussed in several books, e.g., [45] and [111].

1.8 Notes on Literature

Studies of interfacial flows on microscale have been motivated by a number of fascinating practical applications. Examples of such applications, as well as discussion of various aspects of mathematical modeling of microscale flows, can be found in several recently published books on the subject of microfluidics, e.g., Berthier [16], Bruus [25], Kirby [77], and Tabeling [127]. Fundamentals of capillarity and physics of wetting phenomena are discussed in the books of de Gennes et al. [56] and Adamson and Gast [2], as well as a recent review of Bonn et al. [20]. The classical works on the thermocapillary phenomena are reviewed by Davis [42]. More details on the lubrication-type models for thin film flows can be found in Oron et al. [97] and Craster and Matar [40]. Discussion of fluid transport on microscale can be found in review articles by Squires and Quake [120] and Stone et al. [123]. Slip phenomena at solid walls are the subject of the reviews by Lauga et al. [85] and Vinogradova [132]. Disjoining pressure is discussed in Derjaguin et al. [45], Israelachvili [72], and Starov et al. [122].

Chapter 2
Coating Flows and Contact Line Models

2.1 Thin-Film Flow Down an Inclined Plane

2.1.1 Nondimensional Model Formulation

The model of gravity-driven flow formulated in Sect. 1.3 was based on the assumption that the inclined solid surface is initially covered with a liquid film. In experiments, the solid surface is often initially dry and is gradually covered by the liquid film as the leading edge of the film moves down under the action of gravity. Studies of this configuration are important for a number of applications involving coating of a solid surface with a layer of liquid (which is often solidified after the coating is complete). Even though in practical applications on the microscale, such as manufacturing computer hard drives, centrifugal force rather than gravity drives the flow, the key issues in mathematical modeling are essentially the same; we focus on the gravity-driven film flow in the present section.

Consider a film of viscous liquid of density ρ and viscosity μ flowing down a plane inclined at an angle α, as illustrated in Fig. 2.1. As in Sect. 1.3, the flow is assumed to be two-dimensional, with no variations in the direction normal to the plane of the sketch. Far away from the leading edge the film is approximately flat and therefore the flow is described by the constant-thickness solution obtained in Sect. 1.3, with the average flow velocity $\rho g d^2 \sin \alpha / 3\mu$. It is convenient to define the characteristic flow velocity by:

$$U = \frac{\rho g d^2}{\mu} \sin \alpha \tag{2.1}$$

and the capillary number based on this velocity as:

$$Ca = \frac{\mu U}{\sigma}. \tag{2.2}$$

V.S. Ajaev, *Interfacial Fluid Mechanics: A Mathematical Modeling Approach*,
DOI 10.1007/978-1-4614-1341-7_2, © Springer Science+Business Media, LLC 2012

Fig. 2.1 Sketch of
gravity-driven
two-dimensional flow with
a finite contact angle

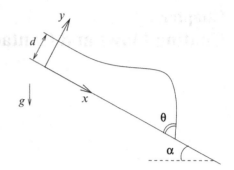

Here, the surface tension σ is assumed constant. At the leading edge of the film, the liquid–gas interface is assumed to be meeting the solid surface at a contact angle θ, as shown in Fig. 2.1. The point of contact between the interfaces in the cross-sectional sketch in Fig. 2.1 corresponds to a line in the actual experiment; this line is moving downward and will be referred to as *a moving contact line*.

Let us now recall the lubrication-type model of a deforming film on a horizontal solid substrate formulated in Sect. 1.4. It was argued there that if the characteristic length scale in the horizontal direction is much larger than the film thickness, equations for liquid flow can be simplified to the system (1.46)–(1.48) or, in nondimensional form, to (1.56)–(1.58). The same arguments can be repeated for the present case of a film on an inclined surface if the length scale of change of film thickness along the plate is much larger than d, except that now gravity has a nonzero projection onto the x-direction, i.e., along the incline. Thus, instead of (1.56)–(1.58) we write (now using subscripts to denote derivatives)

$$p_x = u_{yy} + 1, \tag{2.3}$$

$$p_y = 0, \tag{2.4}$$

$$u_x + v_y = 0. \tag{2.5}$$

Here, x and y are the *nondimensional* Cartesian coordinates scaled by $d/\mathrm{Ca}^{1/3}$ and d, respectively; u is the velocity component in the x-direction scaled by U, v is the velocity component in the y-direction scaled by $\mathrm{Ca}^{1/3}U$, p is difference between the dimensional pressure and the atmospheric pressure scaled by $\mathrm{Ca}^{2/3}\sigma/d$. In Chap. 1 we used tildas to distinguish nondimensional variables from the dimensional ones. In the present chapter, all variables will be assumed nondimensional unless noted otherwise and we will not use tildas. The component of gravity in the y-direction does not appear in our lubrication-type model although it can play an important role in experiments at small values of the inclination angle α. This motivated a slightly different set of assumptions about characteristic scales used, e.g., in Bertozzi and Brenner [17].

The scaled interfacial boundary conditions (1.61)–(1.62) formulated in Sect. 1.4 still apply (except that there is no temperature gradient along the interface) and are written in the notation of the present section as:

$$h_t + uh_x - v = 0 \quad \text{at} \quad y = h(x,t), \tag{2.6}$$

$$u_y = 0 \quad \text{at} \quad y = h(x,t), \tag{2.7}$$

$$p = -h_{xx} \quad \text{at} \quad y = h(x,t), \tag{2.8}$$

where the interface location is described by the function $y = h(x,t)$ and t is nondimensional time, scaled by $\mathrm{Ca}^{-1/3}d/U$.

At the solid boundary, the no-slip condition is satisfied and the normal velocity is zero, so

$$u = 0 \quad \text{at} \quad y = 0, \tag{2.9}$$

$$v = 0 \quad \text{at} \quad y = 0. \tag{2.10}$$

By repeating the same steps as in the derivation of (1.66) in Sect. 1.4, (2.5), (2.6), and (2.10) yield

$$h_t + \left(\int_0^h u\,dy \right)_x = 0. \tag{2.11}$$

Since according to (2.4), p is a function of x and t only, (2.3) can be integrated twice in y to give the following velocity profile,

$$u = \frac{1}{2}(p_x - 1)(y^2 - 2yh). \tag{2.12}$$

Here, we used (2.7) and (2.9) to determine the two constants of integration. By taking into account the boundary condition (2.8), we find

$$u = -\frac{1}{2}(h_{xxx} + 1)(y^2 - 2yh). \tag{2.13}$$

By substituting this velocity profile into (2.11) we obtain an evolution equation for the interface shape in the form

$$h_t + \frac{1}{3} \left(h^3 h_{xxx} \right)_x + h^2 h_x = 0. \tag{2.14}$$

The lubrication-type approach resulting in this equation is applicable to the situation sketched in Fig 2.1 only when the contact angle θ is sufficiently small. In fact, since the contact angle is the inverse tangent of the absolute value of the interface slope at the contact line, the value of θ has to be of the same order of magnitude as $\mathrm{Ca}^{1/3}$, as discussed in more detail in Goodwin and Homsy [58]. From now on, we use the re-scaled contact angle Θ defined by $\Theta = \theta\mathrm{Ca}^{-1/3}$. We assume that Θ is a constant independent of the contact line speed or any other quantities related to the flow and impose the boundary condition

$$h_x(x_{CL},t) = -\Theta, \tag{2.15}$$

where $x_{CL} = x_{CL}(t)$ is the contact-line location which has to be determined as part of the solution in such a way that the condition (2.15) is satisfied and

$$h(x_{CL}, t) = 0. \tag{2.16}$$

Assuming that the film is flat near $x = 0$, the boundary conditions there are

$$h = 1, \quad h_x = 0 \quad \text{at} \quad x = 0. \tag{2.17}$$

The initial condition needed to complete the model formulation is not easy to specify because it depends on the particular choice of the starting procedure in experiments. Fortunately, some results, which turn out to be very important for practical applications, can be obtained using a simplified framework, in which the shape of the interface is assumed stationary (i.e., not changing in time) in a reference frame moving downward with the contact line, at a speed u_{CL}. In this reference frame, the interface is described by a function $h_0(\hat{x})$, $\hat{x} = x - u_{CL}t$ and the substrate is moving with the velocity $-u_{CL}$. The boundary condition (2.9) is then replaced with

$$u = -u_{CL} \quad \text{at} \quad y = 0. \tag{2.18}$$

Near $\hat{x} = 0$, the film is flat and of unit thickness, so the velocity profile satisfying the condition (2.18) together with $u_y = 0$ at the interface is

$$u = -\frac{1}{2}(y^2 - 2y) - u_{CL}. \tag{2.19}$$

The integral of this profile from 0 to 1 has to be zero since otherwise the amount of liquid between $\hat{x} = 0$ and the contact line would change and the interface would not remain stationary. Integrating (2.19) gives the condition $u_{CL} = \frac{1}{3}$, simply reflecting the fact that the contact line speed is equal to the average flow velocity. Now, by using $\hat{x} = x - t/3$ and the chain rule, (2.14) is simplified to an ordinary differential equation in $\hat{x}$:

$$-\frac{1}{3}h_0' + \frac{1}{3}\left(h_0^3 \, h_0'''\right)' + h_0^2 h_0' = 0, \tag{2.20}$$

where primes are used to denote differentiation with respect to $\hat{x}$. This equation can be integrated once, resulting in

$$h_0^2 \, h_0''' = 1 - h_0^2. \tag{2.21}$$

The constant of integration is set to zero based on the condition that in the flat part of the film ($h_0''' = 0$) the value of h_0 is equal to unity.

Equation (2.21) has to be solved on the domain $[0, \hat{x}_{CL}]$ with the contact line coordinate $\hat{x}_{CL}$ chosen large enough to ensure that the liquid film is flat near $\hat{x} = 0$. We choose $\hat{x}_{CL} = 10$ and use the boundary conditions

$$h_0'(0) = 0, \quad h_0(\hat{x}_{CL}) = 0, \quad h_0'(\hat{x}_{CL}) = -\Theta. \tag{2.22}$$

2.1.2 Shear-Stress Singularity

It may seem that the interface shape can now be easily found by solving the boundary value problem defined by (2.21) and (2.22). However, severe convergence problems are encountered if one tries to solve this system numerically, e.g., using the bvp4c solver from MATLAB. To understand the origin of the numerical difficulties let us analyze (2.21) for values of $\hat{x}$ approaching $\hat{x}_{\mathrm{CL}}$. First, we note that since $h_0 \to 0$ as $\hat{x} \to \hat{x}_{\mathrm{CL}}$, (2.21) requires that $h_0''' \to \infty$ near the contact line. Clearly, a solution with this property cannot be accurately described by the standard numerical methods which rely on the assumption that all derivatives of the solution are bounded. Even though some advanced numerical methods can handle situations when the derivatives are unbounded, we will not attempt to solve the boundary value problem defined by (2.21) and (2.22) numerically because solutions with $h_0''' \to \infty$ at the contact line are not physically meaningful: predictions based on such a solution are in obvious contradiction with physical reality. To illustrate this point, let us calculate the shear-stress at the solid–liquid interface based on such a solution. Based on (2.13),

$$u_y|_{y=0} = h(h_{xxx} + 1),\qquad\qquad (2.23)$$

or, in the moving reference frame,

$$u_y|_{y=0} = h_0 \left(h_0''' + 1 \right).\qquad\qquad (2.24)$$

Using (2.21), this equation can be written as:

$$u_y|_{y=0} = h_0^{-1}.\qquad\qquad (2.25)$$

The right-hand side of this equation clearly goes to infinity as $h_0 \to 0$, so the shear-stress is singular at the contact line. The total tangential force acting at the solid–liquid boundary between a point $\hat{x}$ and the contact line is determined by integrating the shear-stress over the area of contact and is proportional to the nondimensional quantity

$$f \equiv \int_{\hat{x}}^{\hat{x}_{\mathrm{CL}}} u_y|_{y=0}\, ds = \lim_{\hat{x} \to \hat{x}_{\mathrm{CL}}^-} \int_{\hat{x}}^{\hat{x}} h_0^{-1} ds.\qquad\qquad (2.26)$$

Here we used (2.25) and the standard definition of an improper integral. If the point $\hat{x}$ is sufficiently close to the contact line, we can approximate local interface shape by a linear function, i.e.,

$$h_0(s) = -\Theta(s - \hat{x}_{\mathrm{CL}}),\qquad \hat{x} \le s \le \hat{x}_{\mathrm{CL}}.\qquad\qquad (2.27)$$

By substituting this into the formula expressing f as a limit, (2.26), we obtain

$$f = -\lim_{\hat{X} \to \hat{x}_{\text{CL}}} \int_{\hat{x}}^{\hat{X}} \frac{\mathrm{d}s}{\Theta(s - \hat{x}_{\text{CL}})} = -\lim_{\hat{X} \to \hat{x}_{\text{CL}}} \frac{1}{\Theta} \ln(\hat{x}_{\text{CL}} - \hat{X}) + \frac{1}{\Theta} \ln(\hat{x}_{\text{CL}} - \hat{x}) = \infty.$$

(2.28)

Thus, our solution predicts an *infinite* value of f, which implies that an infinite force is needed to move the film down an inclined surface, i.e., the film should not move. To provide an even more dramatic illustration of the nonphysical nature of the solution with the shear-stress singularity, consider the following situation. When a small solid object is floating in water, everyday experience suggests that a slight downward push is sufficient to completely immerse it into water. However, if the shear-stress at the moving contact line (where the solid touches the liquid surface) is indeed singular as described above, the resistance to downward motion should be infinite! As Huh and Scriven [68] put it, "not even Herakles could sink a solid if the physical model were entirely valid, which it is not."

While the shear-stress singularity is illustrated here in the context of a simple coating flow, it turns out to be an intrinsic property of moving contact lines, not specific to the particular geometric configuration or to the use of the lubrication theory, as will become evident below in Sect. 2.1.7.

2.1.3 Removing the Singularity

Since the mathematical model developed in Sect. 2.1.1 leads to nonphysical predictions, it is natural to ask how to modify the model so that the physics of the viscous gravity-driven flow down an inclined plane is captured correctly. One such modification, which is particularly easy to implement in our framework, is based on the following experimental observation. When a liquid–gas interface appears to be in contact with the solid, careful experimental studies of the contact line region often indicate that the true contact is not really achieved. Instead, what appears to be dry solid is covered with a very thin film called a *precursor film* [55]. In other words, instead of the true contact between liquid, solid, and gas, as shown in Fig. 2.1, one may be dealing with a situation sketched in Fig. 2.2 (the scaled precursor film thickness b is exaggerated to improve the clarity of the picture).

While moving contact line in the model of Fig. 2.1 is defined precisely by the point at which the film thickness is zero, in the modified model one usually speaks of an *apparent* contact line, defined as a narrow transition region between the thicker film and much thinner precursor. The liquid–gas interface appears to be in contact with the solid there based on observations which do not resolve the precursor film. In some studies, the apparent contact line is defined more precisely by the location at which the interface curvature reaches its local maximum. Assuming that the apparent contact line is moving downward at a constant speed u_{CL}, we consider the moving reference frame such that $\hat{x} = x - u_{\text{CL}}t$ and $\hat{x}$ is in the domain $[0, L]$. We

Fig. 2.2 Sketch of
gravity-driven flow with a
precursor film ahead of the
apparent contact line.
Precursor thickness is
exaggerated for clarity

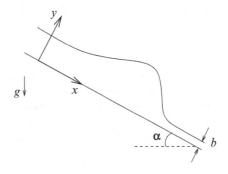

choose L large enough to ensure that the film is flat *both* near $\hat{x} = 0$ and near $\hat{x} = L$; the film thickness near $\hat{x} = L$ is equal to b. The velocity profile near the left endpoint of the domain is still given by (2.19), so the integral of the flow velocity across the film is

$$\int_0^1 u \, dy = \frac{1}{3} - u_{CL}. \tag{2.29}$$

In contrast to the case of true contact line, liquid is now entering the domain at the right endpoint. The velocity profile there is

$$u = -\frac{1}{2}\left(y^2 - 2yb\right) - u_{CL}. \tag{2.30}$$

If the interface, described by $h_0(\hat{x})$, is stationary in our moving reference frame, the total amount of liquid in the domain does not change and the integral of this profile from 0 to b should be equal to (2.29). This condition determines the value of the apparent contact line velocity,

$$u_{CL} = \frac{1 - b^3}{3(1 - b)}. \tag{2.31}$$

Substituting $h = h_0(x - u_{CL}t)$ into the evolution equation (2.14) and using primes to denote derivatives with respect to $\hat{x}$, we obtain

$$-u_{CL}h_0' + \frac{1}{3}\left(h_0^3 \, h_0'''\right)' + h_0^2 h_0' = 0. \tag{2.32}$$

Integration in $\hat{x}$ with the condition of flat film of unit thickness at $\hat{x} = 0$ leads to

$$h_0^2(h_0''' + 1) = 1 + b + b^2 - \frac{b(1 + b)}{h_0}. \tag{2.33}$$

The thickness values are specified at the endpoints,

$$h_0(0) = 1, \quad h_0(L) = b, \tag{2.34}$$

and *all* derivatives have to be zero at both endpoints to ensure that the film is flat there. Clearly, this implies looking for a solution of a third-order equation (2.33) with more than three boundary conditions, so there is no a priori guarantee that such a solution exists. However, careful numerical studies indicate that for sufficiently large computational domain length L the appropriate solution can always be found.

2.1.4 Numerical Methods

The nonlinear boundary value problem formulated in the end of Sect. 2.1.3 can in principle be solved by discretizing the domain $[0, L]$, e.g., in a manner similar to Sect. 1.6, and using the finite-difference approximation of the third derivative to obtain a system of equations for the values of $h_0(\hat{x})$ at the mesh points. The classical Newton's method can then be used to solve the resulting nonlinear system iteratively, as described, e.g., in Kelley [75]. MATLAB codes suitable for iterative solutions of nonlinear systems of equations are available online (see [75] for details). However, these codes often fail to converge for the present problem with uniform mesh discretization. Numerical methods based on nonuniform mesh discretizations can have better convergence properties but they can also fail unless the initial condition is sufficiently close to the final solution. Obtaining such an accurate initial guess of the unknown solution is a challenging task. In recent years, the so-called pseudo-transient continuation methods have been gaining popularity for situations when the Newton's method and its variations have severe convergence problems [34]. The idea is to consider an unsteady extension of the steady state equations and use the method of lines, introduced in Sect. 1.6, to solve the unsteady problem until the solution reaches the steady state within a specified accuracy. Application of this idea to the present situation is especially straightforward since a transient version of (2.33) is readily available in the form of the original evolution equation for the interface shape, (2.14). A traveling-wave solutions of the latter equation was found by relatively minor modifications of the numerical method described in detail in Sect. 1.6. The corresponding MATLAB code can be found in Sect. B.2.1. The numerical method turns out to be quite robust and converges quickly even with the discontinuous initial condition of the form

$$h(\hat{x}, 0) = \begin{cases} 1, & 0 \leq \hat{x} \leq L/2, \\ b, & L/2 < \hat{x} \leq L. \end{cases} \tag{2.35}$$

An alternative approach to solving the boundary value problem from Sect. 2.1.3 is a shooting method developed by Tuck and Schwartz [131]. They use analytical results describing the local behavior of the solution to formulate the boundary conditions at the left endpoint of the computational domain. To illustrate this approach, let us write the solution near $\hat{x} = 0$ in the form

$$h_0(\hat{x}) = 1 + a\zeta(\hat{x}), \tag{2.36}$$

where a is a small parameter and the unknown function ζ can be found by substituting (2.36) into (2.33) and linearizing in a (i.e., neglecting all terms which are nonlinear in a). The equation for ζ obtained by the linearization procedure is

$$\zeta''' + \left(2 - b - b^2\right)\zeta = 0. \tag{2.37}$$

Since $b < 1$, the solution of this linear ordinary differential equation which decays to zero at $-\infty$ can be written in the form

$$\zeta = e^{q\hat{x}}\cos\left(q\sqrt{3}\hat{x}\right), \tag{2.38}$$

where $q = (2 - b - b^2)^{1/3}/2$. The complementary solution involving sine instead of cosine can be eliminated by a shift in the origin of coordinates. By calculating the derivatives of ζ from (2.38) and evaluating the results at $\hat{x} = 0$, we obtain the following boundary conditions at the left endpoint of the computational domain for small a:

$$h_0(0) = 1 + a, \quad h_0'(0) = aq, \quad h_0''(0) = -2aq^2. \tag{2.39}$$

An even more accurate approximation to the solution near $\hat{x} = 0$ can be obtained by incorporating the corrections which are second order in the small parameter a, as was done in [131].

The boundary value problem for the interface shape based on (2.33) can be reformulated in terms of new functions $y_1 = h_0$, $y_2 = h_0'$, and $y_3 = h_0''$, as follows,

$$y_1' = y_2, \tag{2.40}$$

$$y_2' = y_3, \tag{2.41}$$

$$y_3' = \frac{1 + b + b^2}{y_1^2} - \frac{b(1 + b)}{y_1^3} - 1, \tag{2.42}$$

$$y_1(0) = 1 + a, \quad y_2(0) = aq, \quad y_3(0) = -2aq^2, \quad y_1(L) = b. \tag{2.43}$$

The problem can now be solved (with a being a parameter varied as part of the shooting procedure) using standard numerical methods for first-order boundary value problems, e.g., the Matlab solver bvp4c. In practice, the terms which are second-order in a are incorporated in the numerical procedure. The MATLAB code for this method can be found in Sect. B.2.2. More details on using MATLAB solvers for boundary value problems can be found, e.g., in Chap. 3 of [110].

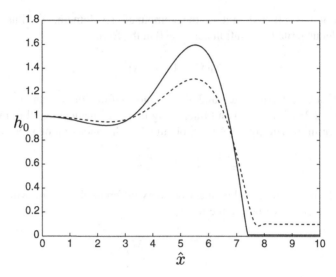

Fig. 2.3 Steady interface shapes in the moving reference frames for $b = 0.01$ (*solid line*) and $b = 0.1$ (*dashed line*)

2.1.5 Interface Shapes

Numerical results for interface shapes $h_0(\hat{x})$ are plotted in Fig. 2.3 for two different values of the scaled precursor thickness, b. The choice of the horizontal coordinate, $\hat{x} = x - u_{CL}t$, implies that the interface shapes in the plot are steady (i.e., independent of time). In terms of the original scaled variables, x and t, these results correspond to traveling wave solutions moving at a speed u_{CL} defined by (2.31). It is clear in Fig. 2.3 that liquid tends to accumulate behind the apparent contact line, forming a bump which is sometimes called the "capillary ridge." By examining the dashed line we also observe that the minimum film thickness is slightly below b but clearly above zero. The same conclusion can be reached for the solid line if one zooms into the apparent contact line region. These observations can be used to clarify why the singularities in the shear-stress and the total resistance force are removed when the precursor film is introduced into the mathematical model. Recall that singularities in integrals such as the one in (2.28) appear only when the film thickness approaches zero and are clearly avoided when the thickness is above a positive number, even if that number is small.

The dimensional precursor film thickness is small compared to the length scale d, but is still assumed to be sufficiently large for the continuum approximation, introduced in Sect. 1.1, to be valid and for the disjoining pressure effects, discussed in Sect. 1.7, to be negligible. It turns out that even when the disjoining pressure defined by (1.129) is incorporated into the model, it does not affect the interface shapes away from the apparent contact line as long as the value of d is above $\sim 1\,\mu$m.

The capillary ridge in Fig. 2.3 is higher for the smaller precursor thickness, but the difference in height is not dramatic despite the order of magnitude difference between the values of b. Let us investigate how the global interface shape depends on the precursor film thickness. In the region where h_0 approaches b, the natural length scale is defined by bd instead of the usual value of d, so (2.33) can be written in terms of new variables $H = h_0/b$ and $X = (\hat{x}_1 - \hat{x})/b$ ($\hat{x}_1$ being an arbitrary point in the apparent contact line region). The result, after neglecting several terms which are small in the limit of small b, is written as follows,

$$H^2 H''' = \frac{1}{H} - 1. \tag{2.44}$$

Here, we use primes to denote derivatives with respect to X. The interface shape at $h_0 \sim 1$ depends on the behavior of the solutions of (2.44) *outside* the contact line region, i.e., in the limit of $X \to \infty$. A solution which grows as X^2 in this limit can be obtained in the form of an infinite series [14], but it is clearly not applicable to our problem because it predicts a very large curvature away from the apparent contact line region. Thus, we follow Bender and Orszag [14] and look for a different asymptotic behavior of the solution at $X \to \infty$, of the form

$$H \sim AX (\ln X)^\beta. \tag{2.45}$$

By substituting this into (2.44), we find $\beta = 1/3$ and $A = 3^{1/3}$. The slope of the interface at large X can then be estimated from the formula

$$H_X = (3 \ln X)^{1/3}. \tag{2.46}$$

In terms of the variables h_0 and $\hat{x}$, this implies

$$h_0'(\hat{x}) = - \left(3 \ln \frac{\hat{x}_1 - \hat{x}}{b} \right)^{1/3}. \tag{2.47}$$

In experiments, the apparent contact angle θ_{app} is typically determined as the absolute value of the slope angle measured at a dimensional distance l^* from the precursor film region, where l^*/d is much larger than b but small compared to $\text{Ca}^{-1/3}$. Based on (2.47) and the definitions of the length scales in the direction along the solid surface and normal to it ($d\,\text{Ca}^{-1/3}$ and d, respectively), we obtain

$$\theta_{\text{app}} = \text{Ca}^{1/3} \left(3 \ln \frac{l^* \text{Ca}^{1/3}}{bd} \right)^{1/3}. \tag{2.48}$$

Since the influence of the factor of $\text{Ca}^{1/3}$ in the logarithmic term on the value of θ_{app} is not very strong, although important for interpreting some experimental data [50],

Fig. 2.4 A sketch of
dynamic contact angle versus
velocity based on
experimental data (after [70])

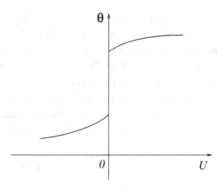

an approximate dependence of θ_{app} on the capillary number is often used instead of
(2.48), in the form

$$\theta_{app} \sim Ca^{1/3}. \tag{2.49}$$

For comparison with experimental data, it is useful to note that the dimensional
contact line speed, U_{CL}, in our model is approximately $U/3$, so (2.48) and (2.49)
can be written as:

$$\theta_{app}^3 = \frac{9\mu U_{CL}}{\sigma} \ln \frac{l^*(3\mu U_{CL})^{1/3}}{\sigma^{1/3}bd} \tag{2.50}$$

and

$$\theta_{app}^3 \sim \frac{\mu U_{CL}}{\sigma}. \tag{2.51}$$

Most experimental data for the case of complete wetting is described by the
approximate formula, (2.51), reasonably well. For the case when the static contact
angle is nonzero, a more general equation, often called Tanner's law, has been used
to describe dynamic contact angle data:

$$\theta^3 = \theta_0^3 + c_T\frac{\mu U_{CL}}{\sigma}, \tag{2.52}$$

where θ_0 is the static contact angle discussed in Chap. 1, c_T is an empirical constant
close to 72. Equation (2.52) is also sometimes referred to as the Hoffman–Voinov–
Tanner law or Cox–Voinov law to recognize the authors of the pioneering studies of
dynamic contact angles [39, 66, 128, 134]. Finally, one feature of the experimental
data not captured by either the precursor film model or the Tanner's law, is that the
contact line can remain stationary for a range of contact angles. Thus, the actual
experimental measurements often result in curves of the type sketched in Fig. 2.4.
The negative velocity corresponds to the case of a receding contact line (discussed
in more detail in Sect. 2.2).

While introduction of the precursor film into the model clearly allows one to
remove the shear-stress singularity, strictly speaking it is only justified when such
film is indeed present. Experiments indicate that this is not always the case and is in

fact unlikely when the observed value of the contact angle is not small. It is natural to ask how the singularity is removed when no precursor film is present. To answer this question, several studies use the notion of slip at the solid surface, introduced in Sect. 1.5*. However, instead of the simple model discussed there, they employ the so-called Navier–Greenspan model of slip, usually written in the form

$$u = \frac{\bar{\alpha}}{3h}\frac{\partial u}{\partial y} \quad \text{at} \quad y = 0, \tag{2.53}$$

where $\bar{\alpha}$ is a constant. The shear-stress singularity is avoided when this condition is used instead of the classical no-slip condition [59]. The value of the contact angle can then be specified in the simulation either as a constant equal to its static value, or as a known function of local velocity, e.g., based on an empirical formula, such as, (2.52) or an experimental curve. Remarkably, the interface shapes away from the contact line region for gravity-driven flow obtained using the Navier–Greenspan model and no precursor film are the same as with the precursor and no slip, as long as $\bar{\alpha} = b$ and both are sufficiently small [119].

An alternative approach to removing the shear-stress singularity has been proposed by Shikhmurzaev and described in detail in [111]. It does not require introduction of either precursor film or slip but rather uses the fact that surface tension is a dynamic quantity which cannot always be assumed equal to its value measured in static configurations. In particular, when an interface is created, the surface tension is initially different from the static value, but relaxes to it over some characteristic time τ_{rel}. Taking this effect into account results in singularity-free models of moving contact lines. However, the approach of Shikhmurzaev remains controversial. Its critics often point out that τ_{rel} used in his model are significantly higher than physically realistic values. More discussion of the issue of the characteristic relaxation times for surface tension can be found in [49, 112].

2.1.6   Stokes Flow: Equations for Stream Function and Vorticity

The models of gravity-driven flow discussed so far have an important limitation: they all are based on the assumption that the characteristic length scale in the direction along the incline is large compared to the film thickness. In experiments, this assumption is violated when films flow over a dry substrate (no precursor) and the contact angle θ (shown in Fig. 2.1) is not small. The Reynolds number $\mathrm{Re} = Ud\rho/\mu$ in many such experiments is still rather small, so the nonlinear terms in the Navier–Stokes equations can be neglected, resulting in the Stokes flow equations introduced in Sect. 1.5*. For the geometric configuration illustrated in Fig. 2.1, these equations can be formulated in nondimensional terms as:

$$p_x = \mathrm{Ca}(\nabla^2 u + 1), \tag{2.54}$$

$$p_y = Ca(\nabla^2 v - \cot\alpha), \qquad (2.55)$$

$$u_x + v_y = 0. \qquad (2.56)$$

Here, $\nabla^2 = \partial_x^2 + \partial_y^2$, all length variables are scaled by d, the velocity components are scaled by U, and pressure is scaled by σ/d. These equations have to be solved with the no-penetration and the Navier–Greenspan slip condition (to avoid shear-stress singularity) at the solid wall and the general interfacial boundary conditions formulated in Sect. 1.5*. For the two-dimensional case considered here, it is often convenient to reformulate the problem in terms of the stream function ψ, defined by:

$$u = \psi_y, \qquad v = -\psi_x. \qquad (2.57)$$

The continuity equation (2.56) is now satisfied automatically. By differentiating equation (2.55) with respect to x and subtracting it from the y-derivative of (2.54), we obtain

$$\nabla^4 \psi = 0. \qquad (2.58)$$

Here, we expressed the velocity components in terms of the stream function. For the purposes of numerical implementation, it is usually more convenient to use an equivalent system of partial differential equations,

$$\nabla^2 \psi = -\omega, \qquad (2.59)$$

$$\nabla^2 \omega = 0, \qquad (2.60)$$

where $\omega = v_x - u_y$ is the vorticity component in the direction normal to the x–y plane (although in some numerical studies ω denotes the quantity $u_y - v_x$). Note that the derivations of (2.58) and (2.59)–(2.60) did not depend on the details of flow geometry, so these equations can be used for a wide range of different flows as long as the Reynolds numbers are small.

There is extensive literature on numerical methods for solving the Stokes flow equations. The details of these methods are beyond the scope of this book. For the geometric configuration shown in Fig. 2.1, numerical steady-state solutions of the Stokes flow equations with the Navier-Greenspan slip condition have been obtained by Goodwin and Homsy [58] using the boundary-element method. Comparison between their numerical results and the lubrication-type approach indicates that the latter works surprisingly well even for relatively large contact angles.

2.1.7 Local Flow Near Moving Contact Line

The shear-stress singularity has been discussed so far only in the context of a simplified model of gravity-driven flow, based on a lubrication-type approach. It is natural to wonder if the singularity may disappear if the characteristic lengths in the x- and y-directions are the same, denoted by d. To illustrate that this is not

Fig. 2.5 Sketch of local
geometry and polar
coordinates near a moving
contact line

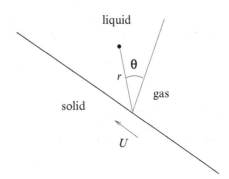

the case and to show that the shear-stress singularity is present in a wide range of
different flows, let us consider local behavior of solutions of fluid flow equations in
the vicinity of a contact line. Even if the flow domain has complicated geometry,
near the contact line the geometry is that of a wedge formed by the liquid–gas
and solid–liquid interfaces, as sketched in Fig. 2.5. It is therefore convenient to
write the Stokes flow equations in terms of polar coordinates (r, θ) shown in the
sketch (r is nondimensional, scaled by d). The value of the contact angle is denoted
by $\hat{\theta}$ here. The radial and angular components of flow velocity, both scaled by a
characteristic flow velocity U, can be expressed in terms of the stream function ψ
in polar coordinates:

$$u_r = \frac{1}{r}\frac{\partial \psi}{\partial \theta}, \quad u_\theta = -\frac{\partial \psi}{\partial r}. \tag{2.61}$$

Assuming the Reynolds number, $\mathrm{Re} = U d \rho / \mu$ is small, the Stokes flow approximation discussed in the previous subsection can be used. Using the well-known
expression for ∇^2 in polar coordinates (see, e.g., Sect. A.6 in the Appendix of [1]),
the Stokes flow equations in terms of the stream function and vorticity, (2.59) and
(2.60), are written as:

$$\frac{1}{r}\frac{\partial}{\partial r}\left(r\frac{\partial \psi}{\partial r}\right) + \frac{1}{r^2}\frac{\partial^2 \psi}{\partial \theta^2} = -\omega, \tag{2.62}$$

$$\frac{1}{r}\frac{\partial}{\partial r}\left(r\frac{\partial \omega}{\partial r}\right) + \frac{1}{r^2}\frac{\partial^2 \omega}{\partial \theta^2} = 0. \tag{2.63}$$

Both ω and ψ are expected to be finite as $r \to 0$ (otherwise, the velocity components
would be infinite, which is clearly nonphysical) and represented by a Taylor series
near $r = 0$. In polar coordinates, this implies local behaviors in the form

$$\psi = r f_1(\theta) + O(r^2), \tag{2.64}$$

$$\omega = r f_2(\theta) + O(r^2), \tag{2.65}$$

where f_1 and f_2 are functions to be determined. In the limit of $r \to 0$, $O(r^m)$ here and below denotes any function $f(r, \theta)$ such that

$$\lim_{r \to 0} \frac{f(r, \theta)}{r^m} \tag{2.66}$$

exists for all θ between zero and $\hat{\theta}$.

By substituting (2.64) and (2.65) into (2.62) and (2.63), we obtain ordinary differential equations for f_1 and f_2,

$$f_1'' + f_1 = -f_2, \tag{2.67}$$

$$f_2'' + f_2 = 0, \tag{2.68}$$

and thus

$$f_1 = A \sin \theta + B \cos \theta + C \theta \sin \theta + D \theta \cos \theta. \tag{2.69}$$

The constants A, B, C, and D are found from the boundary conditions of zero normal velocity and zero tangential stress at the liquid–gas interface,

$$\frac{\partial \psi}{\partial r}(r, 0) = 0, \qquad \frac{\partial^2 \psi}{\partial \theta^2}(r, 0) = 0, \tag{2.70}$$

and the conditions of zero normal and specified tangential velocity at the moving (in the reference frame of the contact line) solid wall,

$$\frac{\partial \psi}{\partial r}(r, \hat{\theta}) = 0, \qquad -\frac{1}{r}\frac{\partial \psi}{\partial \theta}(r, \hat{\theta}) = u_{\text{CL}}. \tag{2.71}$$

From the conditions at the liquid–gas interface (2.70), we immediately obtain $B = 0$ and $C = 0$. The remaining two conditions (2.71) result in a linear system of algebraic equations for the coefficients A and D. By solving this system, we obtain

$$A = u_{\text{CL}} \left(\frac{\sin \hat{\theta}}{\hat{\theta}} - \frac{1}{\cos \hat{\theta}} \right)^{-1}, \tag{2.72}$$

$$D = u_{\text{CL}} \left(\frac{\hat{\theta}}{\sin \hat{\theta}} - \cos \hat{\theta} \right)^{-1}. \tag{2.73}$$

Thus, all unknown constants have been found and therefore the behavior of the components of the stress tensor $\mathbf{T}$ at the solid–liquid interface can be determined. The tangential component of the stress acting in the liquid near the solid wall is expressed in terms of the unit vectors of the polar coordinate system as $\mathbf{e}_r \cdot \mathbf{T} \cdot \mathbf{e}_\theta$, which leads to the conclusion that shear-stress is determined by:

$$T_{r\theta} = -\frac{2\text{Ca}}{r} D \sin \hat{\theta} + O(1). \tag{2.74}$$

Thus, the shear-stress is proportional to r^{-1}, i.e., has exactly the singularity predicted by our simplified lubrication-type model. We note that pressure can be shown to have the same r^{-1} singularity [68].

2.2 Landau–Levich Problem

Moving contact lines discussed throughout most of the previous section are the so-called advancing contact lines: they advance into a region which is either dry or covered with a very thin precursor film. In many practical applications, the receding contact lines are also important, corresponding to the case when the moving interface leaves behind a region of the solid surface which is either dry or covered with a thin liquid film. The thickness of the latter is a function of the contact line speed and can be significantly higher than typical thickness values for the precursor films discussed in Sect. 2.1. The purpose of the present section is to discuss a model of the film left behind a moving interface using a simple configuration relevant for coating applications. A number of interesting phenomena observed when the solid surface behind a moving fluid interface is dry, i.e., the case of the "true" receding contact line, is beyond the scope of the present book but is discussed, e.g., in the review article of Bonn et al. [20].

Let us recall the static configuration of an air–liquid interface near a vertical solid wall discussed in Sect. 1.2 and sketched in Fig. 1.2. Now suppose that instead of being stationary the solid wall is moving upward at a speed U, as shown in Fig. 2.6, a situation relevant for dip coating and other similar practical applications. The problem of determining the interface shape for this geometric configuration, often referred to as the Landau–Levich problem, has been solved in [83]. The first step in the solution is developing a model of viscous flow in the liquid generated as a result of wall motion. Assuming steady flow at small values of the Reynolds number, we can use the Stokes flow equations, discussed in Sect. 1.5*. A nondimensional vector form of these equations is

$$\nabla p = \text{Ca} \nabla^2 \mathbf{u} + \mathbf{e}_g, \tag{2.75}$$

$$\nabla \cdot \mathbf{u} = 0. \tag{2.76}$$

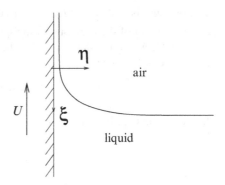

Fig. 2.6 Sketch of the geometry of the Landau–Levich problem

Here, $\mathbf{e}_g$ is the unit vector in the direction of gravity, the components of the velocity vector $\mathbf{u}$ are scaled by U, length variables by $a = \sqrt{\sigma/\rho g}$, and pressure by σ/a. The capillary number $\mathrm{Ca} = \mu U/\sigma$ is typically small, suggesting that the mathematical description of the flow can be simplified further. The previously discussed lubrication-type models of thin films were based on the choices of scales which effectively made the left-hand side and the first term on the right-hand side of (2.75) equally important in the limit of $\mathrm{Ca} \to 0$. Since the pressure gradients are determined by the capillary pressure jump at the interface, this mathematical formulation is equivalent to stating that viscous effects and surface tension both contribute to the shape of the interface. There is no reason to expect this to be the case everywhere in the configuration shown in Fig. 2.6. Far away from the slowly moving wall, the interface shape is not likely to be affected by the wall motion and is determined by the combined action of the capillary forces and gravity. The effects of viscosity and capillarity balance each other only near the wall, in a localized transition region of yet unknown characteristic dimensions L_ξ (along the wall) by L_η (normal to the wall). Let us describe the solution in this region using stationary Cartesian coordinates ξ and η shown in Fig. 2.6, scaled by L_ξ and L_η, respectively. The air-liquid interface is represented by a function $\eta = H(\xi)$ measuring the distance between the interface and the wall. Note that for sufficiently small slope of the interface, the capillary pressure jump is proportional to $H''(\xi)$ or, in dimensional terms, to $\sigma L_\eta/L_\xi^2$. Since the solution in the transition region and the static interface shape found in Sect. 1.2 have to be matched asymptotically in the limit of $\mathrm{Ca} \to 0$, the characteristic capillary pressure jump in the transition region should be independent of the wall velocity. Therefore,

$$\frac{L_\eta}{L_\xi^2} \sim \text{const.} \quad \text{as} \quad \mathrm{Ca} \to 0. \tag{2.77}$$

The capillary pressure gradient in the ξ-direction has to balance the effects of viscosity in the corresponding Stokes flow equation, which can now be written in nondimensional form as:

$$\frac{L_\eta}{L_\xi^3} p_\xi = \mathrm{Ca} \left(\frac{1}{L_\xi^2} u_{\xi\xi} + \frac{1}{L_\eta^2} u_{\eta\eta} \right) + \frac{\rho g}{\sigma}. \tag{2.78}$$

Here, pressure is scaled by $\sigma L_\eta/L_\xi^2$, based on the estimate of the capillary pressure jump in the end of the previous paragraph, and the velocity u in the ξ-direction is scaled by U. Assuming that $L_\eta \ll L_\xi$, the capillary and viscous terms in (2.78) are of the same order when

$$\frac{L_\eta}{L_\xi} \sim \mathrm{Ca}^{1/3}. \tag{2.79}$$

This condition together with (2.77) leads to

$$L_\xi \sim \mathrm{Ca}^{1/3}, \quad L_\eta \sim \mathrm{Ca}^{2/3}. \tag{2.80}$$

From now on, we *define* the scales L_ξ and L_η to be equal to $a\mathrm{Ca}^{1/3}$ and $a\mathrm{Ca}^{2/3}$, respectively. Using the same argument as in Sect. 1.4.2, we conclude that the ratio of the velocity scales in the two directions considered should be proportional to $\mathrm{Ca}^{1/3}$, so the velocity component in the η-direction, v, is scaled by $\mathrm{Ca}^{1/3}U$. By rewriting the standard dimensional Stokes flow equations from Sect. 1.5* in terms of our nondimensional variables and neglecting all terms which are small in the limit of $\mathrm{Ca} \to 0$, we obtain

$$p_\xi = u_{\eta\eta}, \tag{2.81}$$

$$p_\eta = 0, \tag{2.82}$$

$$u_\xi + v_\eta = 0. \tag{2.83}$$

Applying similar nondimensionalization procedure to the standard dimensional boundary conditions formulated in Sect. 1.5* and taking the limit of $\mathrm{Ca} \to 0$, we find

$$uH' - v = 0 \quad \text{at} \quad \eta = H(\xi), \tag{2.84}$$

$$u_\eta = 0 \quad \text{at} \quad \eta = H(\xi), \tag{2.85}$$

$$p = -H'' \quad \text{at} \quad \eta = H(\xi). \tag{2.86}$$

Here, we assume that the interface is stationary (defined by the function $H(\xi)$) and use primes to denote derivatives with respect to ξ. If there is no slip at the moving solid wall, the nondimensional boundary conditions there are written in the form

$$u = -1 \quad \text{at} \quad \eta = 0, \tag{2.87}$$

$$v = 0 \quad \text{at} \quad \eta = 0. \tag{2.88}$$

Integrating (2.83) in η with the conditions (2.84) and (2.88) leads to the integral mass balance of the form

$$\frac{\mathrm{d}}{\mathrm{d}\xi} \int_0^H u\,\mathrm{d}\eta = 0, \tag{2.89}$$

which simply states that the flow rate in each cross-section normal to the ξ-axis is the same. The velocity profile $u(\eta)$ found by integrating (2.81) with the conditions (2.85) and (2.87) is

$$u = \frac{1}{2}p_\xi(\eta^2 - 2\eta H) - 1. \tag{2.90}$$

Substitution of this profile into (2.89) with subsequent integration in η leads to a differential equation for $H(\xi)$:

$$\left(\frac{H^3}{3}H''' - H\right)' = 0. \tag{2.91}$$

Integrating both sides in ξ and rearranging terms gives

$$H''' = \frac{3(H - b)}{H^3}. \tag{2.92}$$

Here, we introduced a constant b. Since the film is nearly flat for $\xi \to -\infty$, the constant b defines the scaled thickness of the film there and the following boundary conditions can be applied,

$$H(-\infty) = b, \quad H'(-\infty) = H''(-\infty) = 0. \tag{2.93}$$

Since b in unknown, an additional boundary condition is needed to determine the interface shape. It is obtained from the condition of matching to the local curvature of the static interface shape found in Sect. 1.2. According to (1.8), the dimensional curvature of the interface is equal to $\rho g h / \sigma$. Near the wall, the static solution reaches its maximum height determined by (1.10) and equal to $a\sqrt{2}$ for the case of perfect wetting. Thus, the dimensional curvature near the wall is $\sqrt{2}/a$, which in nondimensional terms gives the following matching condition

$$H''(\infty) = \sqrt{2}. \tag{2.94}$$

This completes our formulation of the boundary value problem for $H(\xi)$. However, developing a numerical solution procedure becomes more straightforward if (2.92) is written in the form independent from the unknown constant b using new variables

$$\hat{H} = \frac{H}{b}, \quad \hat{\xi} = \frac{3^{1/3}\xi}{b}. \tag{2.95}$$

In terms of these new variables, the interface shape is described by:

$$\hat{H}''' = \frac{\hat{H} - 1}{\hat{H}^3}, \tag{2.96}$$

$$\hat{H}(-\infty) = 1, \quad \hat{H}'(-\infty) = \hat{H}''(-\infty) = 0, \quad \hat{H}''(\infty) = \frac{b\sqrt{2}}{3^{2/3}}, \tag{2.97}$$

where the primes now denote derivatives with respect to $\hat{\xi}$. Suppose the origin of coordinates is chosen such that the solution near $\hat{\xi} = 0$ is close to $\hat{H} = 1$ and can therefore be represented by:

$$\hat{H} = 1 + \zeta, \quad |\zeta| \ll 1. \tag{2.98}$$

By substituting this expression into (2.96) and linearizing in ζ, we obtain

$$\zeta''' = \zeta. \tag{2.99}$$

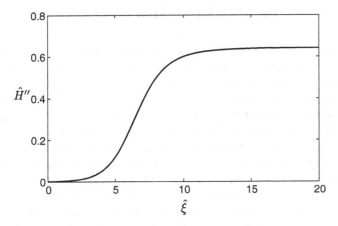

Fig. 2.7 Second derivative of the nondimensional interface position for the Landau–Levich problem, found by solving (2.96) with the conditions at $\hat{\xi} = 0$ given by (2.101), $\delta = 10^{-3}$

Solving this equation with the condition of $\zeta(-\infty) = 0$ leads to the formula

$$\zeta = \delta e^{\hat{\xi}}, \tag{2.100}$$

where δ is a constant. Based on (2.98), the values of the function $\hat{H}$ and its derivatives at $\hat{\xi} = 0$ can then be expressed in terms of δ as follows,

$$\hat{H}(0) = 1 + \delta, \quad \hat{H}'(0) = \hat{H}''(0) = \delta. \tag{2.101}$$

Equation (2.96) is solved numerically with these conditions forward in $\hat{\xi}$ until $\hat{H}''(\hat{\xi})$ flattens out or, equivalently, $\hat{H}'''(\hat{\xi})$ approaches zero. We use $\delta = 10^{-3}$, although any small value can be chosen here since changing δ is equivalent to shifting the origin of coordinates (as long as $\delta \ll 1$). The MATLAB code for this problem can be found in Sect. B.3. Figure 2.7 illustrates the behavior of the second derivative of the interface shape obtained from the numerical solution, showing that at $\hat{\xi} \sim 10$ the derivative is already close to its asymptotic value, $\hat{H}''(\infty)$. Once the latter is obtained with the desired accuracy, the scaled trailing film thickness can be found based on $b = 3^{2/3}\hat{H}''(\infty)/\sqrt{2}$, resulting in the value of $b = 0.9458$. In dimensional terms, the final result for the thickness of the liquid film deposited on the wall can then be written as:

$$H^* = 0.9458\, a\, \mathrm{Ca}^{2/3}. \tag{2.102}$$

This result has been verified experimentally as discussed, e.g., in [88].

2.3 Axisymmetric Spreading of Thin Droplets

Both receding and advancing moving contact lines appear in a wide variety of applications. Understanding the behavior of dynamic contact angle as a function of local velocity is essential for correctly predicting the evolution of fluid interfaces seen in experiments. As a simple illustration, consider spreading of an axisymmetric droplet of a completely wetting liquid on a horizontal solid surface. The geometry of the problem is shown in Fig. 2.8. In the absence of evaporation and condensation, the droplet volume is constant during spreading, so it is convenient to take the cube root of the volume as the scale for all length variables. For small droplets, the Bond number (discussed in Sect. 1.2) is small, so we can neglect the effects of gravity. Since the changes in droplet shape are due to capillary forces only, this regime is often called "capillary spreading". For relatively slow spreading considered here, one can assume that the droplet goes through a sequence of equilibrium shapes, which in the absence of gravity are the shapes of constant-curvature. For thin droplets, they are defined by:

$$h(r) = \frac{2}{\pi R^2} \left(1 - \frac{r^2}{R^2} \right), \tag{2.103}$$

where R is the scaled droplet radius, i.e., the distance from the axis of symmetry to the contact line, as shown in Fig. 2.8.

For each constant-curvature shape, the contact angle is nonzero, so none of these shapes represent a true equilibrium. The latter would require the contact angle to be equal to its static value, which for a completely wetting liquid considered here is zero. As discussed in Sect. 2.1, the apparent contact angle θ_{app} for a contact line moving at a speed U_{CL} can be approximated by:

$$\theta_{app} \sim \left(\frac{\mu U_{CL}}{\sigma} \right)^{1/3}. \tag{2.104}$$

In our nondimensional variables, this implies

$$\theta_{app} \sim (R'(t))^{1/3}, \tag{2.105}$$

where the time scale is determined by the length scale divided by U_{CL}. Equation (2.103), on the other hand, allows us to approximate the apparent contact angle by

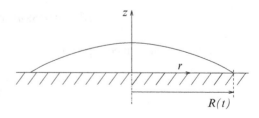

Fig. 2.8 Sketch of axisymmetric spreading droplet and cylindrical coordinates

the absolute value of $h'(r)$ at $r = R$, so that

$$\theta_{app} = \frac{4}{\pi R^3}.$$ (2.106)

By combining (2.105) and (2.106), we immediately obtain $R'(t) \sim R^{-9}$ and therefore

$$R(t) \sim t^{1/10}.$$ (2.107)

This power law has been indeed observed in several experiments, as discussed in more detail, e.g., in [20] and [97].

The solution above predicts that the droplet radius will grow indefinitely, although at a rapidly reducing rate. It may seem natural to assume that the spreading will continue until the continuum approximation breaks down, e.g., when droplet thickness is down to molecular dimensions. However, de Gennes [55] argued that the macroscopic model of spreading actually breaks down at much larger droplet thickness, when the effects of disjoining pressure (introduced in Sect. 1.7) become important and can result in an equilibrium pancake-type shape of the droplet. The thickness of such droplet is much larger than the molecular dimensions at which the continuum approximation breaks down.

It is important to emphasize that the power law given by (2.107) was derived based on the assumption of small Bond number, Bo $\ll$ 1, so it breaks down when the Bond number is close to unity and the effects of gravity become comparable to those of surface tension. In the limit of Bo $\gg$ 1, spreading is dominated by gravity rather than surface tension and a different power law for the radius $R(t)$ can be obtained, of the form

$$R(t) \sim t^{1/8},$$ (2.108)

as discussed in detail, e.g., in Chap. 6 of Leal [87]. This result is obtained by considering the effects of gravity and viscous dissipation of energy in the bulk of the droplet. If most of the dissipation takes place near the contact line, the spreading law is in the form

$$R(t) \sim t^{1/7}.$$

Comparison between different spreading laws and experimental data is discussed by Bonn et al. [20], Oron et al. [97], and Starov et al. [122].

2.4* Fingering Instability of Gravity-Driven Flow

2.4.1 Physical Phenomena

Gravity-driven flow of viscous liquid down an inclined plane can be described by two-dimensional models, such as the ones discussed in Sect. 2.1, only when the contact line remains straight as it moves down the incline. In experiments [69, 114],

Fig. 2.9 A sketch illustrating the shapes of deformed apparent contact line for patterns of wedge-shaped fingers (*top*) and rivulets (*bottom*) for flow down an incline. A view from the top is shown and liquid is above the contact line. Dashed lines show the contact line positions at a later moment in time. After [114]

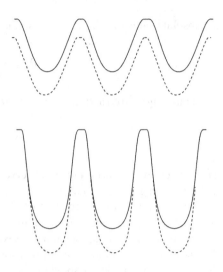

even if the contact line is initially straight, it quickly deforms, resulting in the formation of finger-like structures. The instability leading to this pattern is referred to as *fingering instability* (although the same term is used for the instability of an interface moving in a narrow gap between two parallel plates or in a porous material [67]). Alternatively, the fingering instability of the gravity-driven flow has been referred to as the "rivulet instability." Typical patterns seen in experiments on gravity-driven thin-film flow are illustrated by the sketches in Fig. 2.9. In Fig. 2.9 (top), the pattern consists of wedge-shaped fingers, while Fig. 2.9 (bottom) illustrates elongated fingers with nearly straight contact line on the sides; the latter are also often referred to as rivulets. Both patterns can be observed at the same inclination angle using different liquids. For example, Huppert [69] recorded wedge-shaped fingers for silicone oil and rivulets for glycerine for the inclination angle of $\alpha = 12°$; note that the apparent contact angle was higher for glycerine. Similar fingering phenomena are observed in the process of spin coating in which the centrifugal force drives the flow of liquid over a solid surface. Formation of fingering pattern can result in poor quality of coating, so theoretical studies of fingering instability are important for practical applications in which spin coating process is used, e.g., in semiconductor industry.

The main mechanism of the instability has been elucidated by Spaid and Homsy [119]. It can be explained as follows. Suppose, the contact line is a straight line moving downward at a constant velocity and the height of the capillary ridge is uniform along the direction of the contact line. Recall that the capillary ridge is a common feature of two-dimensional interface shapes, formed as a result of accumulation of liquid behind the apparent contact line, as seen, e.g., in Fig. 2.3. Suppose a small localized bump is formed on top of the capillary ridge as a result of random fluctuations. Then the viscous resistance to the flow in the bump region will be less due to increased local thickness of the film there and the bump will move downward faster than the parts of the capillary ridge away from it, leading to the fingering instability.

The above explanation of the instability mechanism is, of course, only qualitative. In order to develop a quantitative model of the phenomenon, it is natural to employ the ideas of lubrication-type approach since the film thickness is typically small compared to all other macroscopic length scales observed in the system. However, the evolution equation derived in Sect. 2.1 is not directly applicable to the present situation since the assumption of two-dimensionality (film thickness being a function of time and the x-variable) is no longer applicable. Therefore, our first step is to generalize the lubrication-type approach used in Sect. 2.1 to a situation when the liquid–gas interface is a two-dimensional surface and the liquid flow is three-dimensional.

2.4.2 Lubrication-Type Equations in 3D

The key steps in the derivation presented in this subsection are essentially the same for a wide range of problems including fingering instability, nonaxisymmetric spreading of thin liquid droplets, and growth of dry patches in thin liquid films. These steps are illustrated here in the context of fingering instability for gravity-driven flow down a flat solid plate inclined at an angle α. The film far upstream from the contact line is assumed flat and of thickness d, so the characteristic flow velocity U and the capillary number Ca are still defined by (2.1) and (2.2). The formulation is very similar to the one used in the first section of this chapter except that we no longer invoke the assumption of two-dimensional flow and use a different notation for the scaled Cartesian coordinates: the z-coordinate measures the distance in the direction normal to the plate, scaled by d, while the xy plane coincides with the solid surface. Both x and y are scaled by $d/\mathrm{Ca}^{1/3}$, and the x-axis is in the direction of fastest descent along the inclined surface. The lubrication-type equations for the flow are

$$p_x = u_{zz} + 1, \tag{2.109}$$

$$p_y = v_{zz}, \tag{2.110}$$

$$p_z = 0, \tag{2.111}$$

$$u_x + v_y + w_z = 0. \tag{2.112}$$

Here u and v are velocities in the x- and y-directions, respectively, both scaled by U. The velocity in the z-direction, w, is scaled by $\mathrm{Ca}^{1/3}U$, and the pressure p is scaled by $\mathrm{Ca}^{2/3}\sigma/d$.

At the two-dimensional interface, defined by the function $z = h(x,y,t)$, we use the nondimensional versions of the general interfacial conditions formulated in Sect. 1.5*. In the limit of small capillary numbers, these nondimensional conditions simplify to

$$h_t + uh_x + vh_y - w = 0 \quad \text{at} \quad z = h(x,y,t), \tag{2.113}$$

$$u_z = 0 \quad \text{at} \quad z = h(x,y,t), \tag{2.114}$$

$$v_z = 0 \quad \text{at} \quad z = h(x,y,t), \tag{2.115}$$

$$p = -\nabla^2 h \quad \text{at} \quad z = h(x,y,t), \tag{2.116}$$

where we use the operator $\nabla = (\partial/\partial x, \partial/\partial y)$ and the time variable scaled by $Ca^{-1/3}d/U$.

At the solid boundary, the no-slip condition is satisfied and the normal velocity is zero, so

$$u = v = w = 0 \quad \text{at} \quad y = 0. \tag{2.117}$$

By repeating the same steps as in the derivation of (2.14), the following evolution equation is obtained

$$h_t + \frac{1}{3}\nabla \cdot \left(h^3 \nabla \nabla^2 h \right) + h^2 h_x = 0. \tag{2.118}$$

As with the two-dimensional version of this problem discussed in Sect. 2.1, it is convenient to introduce a moving reference frame by defining

$$\hat{x} = x - u_{CL}t, \quad u_{CL} = \frac{1}{3}(1 + b + b^2), \tag{2.119}$$

where b is the scaled thickness of the precursor film, introduced in Sect. 2.1.3. In the moving reference frame, (2.118) can be written as:

$$h_t - \frac{1}{3}(1 + b + b^2)h_{\hat{x}} + \frac{1}{3}\hat{\nabla} \cdot \left(h^3 \hat{\nabla}\hat{\nabla}^2 h \right) + h^2 h_{\hat{x}} = 0, \tag{2.120}$$

where $\hat{\nabla} = (\partial/\partial\hat{x}, \partial/\partial y)$. Far upstream and downstream from the apparent contact line the film is assumed to be flat, with the local thickness values specified as:

$$h(-\infty, y, t) = 1, \quad h(\infty, y, t) = b. \tag{2.121}$$

2.4.3 Linear Stability

Equation (2.120) has a two-dimensional steady-state solution $h = h_0(\hat{x})$ found in Sect. 2.1. However, in order for this solution to be observed in experiments, it has to be stable with respect to small perturbations. Stability conditions can be obtained using the linear stability theory. Let us introduce a small perturbation of the steady-state solution,

$$h(\hat{x}, y, t) = h_0(\hat{x}) + \zeta(\hat{x}, y, t), \quad |\zeta| \ll |h_0|. \tag{2.122}$$

Substituting this form of the solution into (2.120) and neglecting terms which are nonlinear in the small perturbation ζ, we obtain a linearized equation for the perturbation:

$$\zeta_t - \frac{1}{3}(1 + b + b^2)\zeta_{\hat{x}} + \frac{1}{3}\hat{\nabla} \cdot \left(h_0^3 \hat{\nabla}\hat{\nabla}^2 \zeta \right) + \left(h_0^2 h_0''' \zeta \right)_{\hat{x}} + h_0^2 \zeta_{\hat{x}} + 2h_0 h_0' \zeta = 0, \tag{2.123}$$

where primes are used to denote derivatives with respect to $\hat{x}$. Due to linearity of the equation for ζ, any perturbation which is periodic in the y-direction can be represented as a superposition of perturbations of the form

$$\zeta = G(\hat{x})e^{\gamma t + iky}, \tag{2.124}$$

where γ and k denote the growth rate and the wavenumber of the perturbation, respectively, both expressed in nondimensional terms; $G(\hat{x})$ is a function to be determined as part of the solution. Note that ζ is not periodic in the $\hat{x}$-direction due to the conditions of flat film far upstream and downstream from the apparent contact line, so the function $G(\hat{x})$ cannot be represented as a superposition of sinusoidal perturbations in the $\hat{x}$-direction. The stability problem is now reduced to finding an expression for the perturbation growth rate γ as a function of its wavenumber k. Since the extent of the domain in both positive and negative y-directions is infinite, perturbations of all wavenumbers should be considered in the analysis. By substituting the form of the perturbation (2.124) into (2.123), we obtain

$$\gamma G - \frac{1}{3}(1+b+b^2)G' + \left[\frac{h_0^3}{3}(G''' - k^2 G') + h_0^2(h_0''' + 1)G \right]' + \frac{h_0^3}{3}(k^4 G - k^2 G'') = 0. \tag{2.125}$$

The perturbation does not affect the solution far upstream from the apparent contact line or in the precursor film, so the function $G(\hat{x})$ satisfies the conditions

$$G(-\infty) = G'(-\infty) = G(\infty) = G'(\infty) = 0. \tag{2.126}$$

Equations (2.125) and (2.126) define an *eigenvalue problem*, meaning that the nontrivial solutions exist only for certain values of γ. To solve this eigenvalue problem numerically, we consider (2.125) on a finite domain $[0, L]$, assuming that G is zero outside of this domain and satisfies

$$G(0) = G'(0) = G(L) = G'(L) = 0. \tag{2.127}$$

We discretize the domain $[0, L]$ using a uniform mesh and apply the standard finite difference formulas for the derivatives of G. The function h_0 and its derivatives are evaluated based on the numerical solutions obtained in Sect. 2.1 using the code from Sect. B.2.1 and illustrated in Fig. 2.3. The discretized version of the problem then takes the form

$$\gamma \mathbf{G} = A\mathbf{G}, \tag{2.128}$$

where A is a known matrix which depends on h_0 and the mesh size, $\mathbf{G}$ is the vector of values of the function G at the mesh points. The system can now be solved using standard numerical methods for finding eigenvalues of matrices, e.g., in MATLAB. The value of γ with the largest real part determines stability and is shown as a function of the wavenumber in Fig. 2.10 for different values of the precursor film thickness. Based on the numerical solution, the system is always unstable.

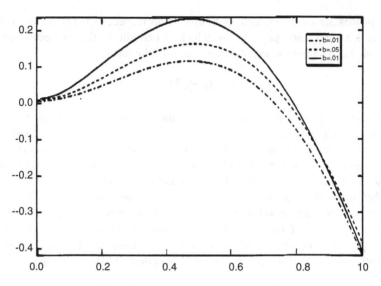

Fig. 2.10 A plot of perturbation growth rate versus the wavenumber based on the numerical solution of (2.125). From [119], reprinted with permission of AIP

A simple plot of different terms in the expression for γ shows that the terms corresponding to gravity effect (and not various terms corresponding to surface tension effects) are most important for the instability, thus justifying the explanation of the instability mechanism discussed in the beginning of this section. This argument can be made more rigorous by considering the rate of energy dissipation as was done by Spaid and Homsy [119].

2.4.4 Numerical Simulations

The linear stability analysis only describes evolution of small perturbations of the two-dimensional solution, so it is not applicable to modeling the interface shapes seen, e.g., in Fig. 2.9. In order to describe experimentally observed fingering patterns, (2.120) has to be solved numerically. For the numerical simulations, a finite size domain has to be chosen. We consider (2.120) on a rectangular domain

$$0 \leq \hat{x} \leq L, \quad 0 < y < \tilde{L}. \tag{2.129}$$

and use the boundary conditions

$$h(0,y,t) = 1, \quad h(L,y,t) = b \tag{2.130}$$

together with the no-flux conditions as described in [80]. Since the experimental fingering patterns such as the ones seen in Fig. 2.9 are usually nearly periodic, we use the periodic boundary conditions at $y = 0$ and $y = \tilde{L}$.

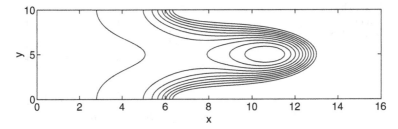

Fig. 2.11 A wedge-shaped finger obtained from numerical simulation of fingering instability for the case of perfect wetting

Different approaches have been used for numerical simulations of fingering. Many of them are based on ADI (alternating direction implicit), an efficient method which allows one to track evolution of two-dimensional interfaces without having to deal with very large matrices [140]. However, recent progress in matrix computations suggests that even a more straightforward method, a version of the method of lines with iterative approaches to matrix inversion is suitable for numerical studies of fingering. Mesh refinement near the contact line, where the interface slope changes most rapidly, makes the simulations more efficient; evaluation of derivatives in this framework is discussed in the following subsection. Numerical simulations have been conducted for both precursor model [51, 80] and the Navier–Greenspan slip model [94], and are in agreement with the linear stability results when the deviation of the interface shape from the solution described by $h_o(\hat{x})$ is small. Some issues have been raised regarding the validity of the linear theory at small inclination angles, and are attributed to the non-self-adjoint nature of the linear operator involved, as discussed by Bertozzi and Brenner [148] and Grigoriev [60].

Numerical simulations of strongly nonlinear evolution of the interface (outside the range of validity of the linear theory) allowed several groups [51, 80, 94] to obtain two different shapes of fingers found in experiments (and shown in Fig. 2.9) and identify criteria for the transition between them. Here, the discussion is based on the work of Moyle et al. [94], in which the contact angle θ is a parameter of the problem and the Navier–Greenspan condition is used to eliminate the shear-stress singularity. For the values of θ equal or close to zero, the pattern is that of wedge-shaped fingers and is steady in a moving reference frame. An example of a wedge-shaped finger seen in numerical simulations is shown in Fig. 2.11 and can be compared to the sketch in Fig. 2.9(top). As the contact angle is increased, fingers become more elongated and eventually transition to rivulet-type shapes. From a practical point of view, it is most important to identify the transition to straight-sided rivulets such that the front of the pattern advances but the troughs between the rivulets remain stationary. This type of pattern results in failure to completely coat the solid surface with a layer of liquid. Complete coating is usually the goal in practical applications. Moyle et al. [94] observe that in the numerical simulation the transition takes place at

$$(\mu U/\sigma)^{-1/3}\tan\theta \approx 1, \tag{2.131}$$

where U is defined by (2.1).

Fig. 2.12 Sketch of a
nonuniform mesh in the
vicinity of a mesh point i

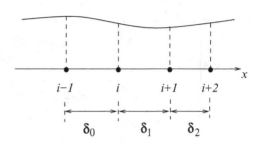

2.4.5   Finite-Difference Method with Nonuniform Mesh

In order to develop efficient numerical methods for problems in which the derivatives of the interface shape become large in localized regions in space, one has to consider using nonuniform meshes. The goal of the present section is to provide brief derivations of the basic finite-difference formulas for derivatives on nonuniform meshes.

Consider an arbitrary nonuniform mesh as sketched in Fig. 2.12. We assume that as the mesh is refined, the distances between mesh points change according to $\delta_k = A_k \delta$, where the constants A_k do not change. As a simple illustration of the method, let us evaluate the second derivative of the interface at the mesh point i in terms of values of h at points in its vicinity. Using Taylor series,

$$h_{i+1} = h_i + h'\delta_1 + \frac{1}{2}h''\delta_1^2 + O(\delta^3), \tag{2.132}$$

$$h_{i-1} = h_i - h'\delta_0 + \frac{1}{2}h''\delta_0^2 + O(\delta^3), \tag{2.133}$$

where h' and h'' are the (unknown) exact values of the derivatives at the point i. The notation $O(\delta^3)$ refers to behavior in the limit of $\delta \to 0$, which is appropriate to consider since δ becomes small as the mesh is refined.

Dividing (2.132) by δ_1, (2.133) by δ_0, and adding the two together allows us to eliminate h'. Then, h'' is expressed as:

$$h'' = \frac{2}{\delta_0 + \delta_1}\left(\frac{h_{i+1} - h_i}{\delta_1} + \frac{h_{i-1} - h_i}{\delta_0}\right) + O(\delta). \tag{2.134}$$

The same approach can be used to express higher-order derivatives except that more mesh points are needed.

Derivatives at points halfway between the mesh points are also often needed for numerical schemes. Let us derive a formula for the third derivative of the interface shape at $i + \frac{1}{2}$ (half-way between the points i and $i+1$). Taylor series can be used to express h_{i+1} in terms of the values of the function h and its derivatives (h', h'', and h''') at $i + \frac{1}{2}$ as follows,

$$h_{i+1} = h + h'\frac{\delta_1}{2} + h''\frac{\delta_1^2}{8} + h'''\frac{\delta_1^3}{48} + O(\delta^4). \tag{2.135}$$

Similar expansions can be written for h_i, h_{i+2}, and h_{i-1}. When the $O(\delta^4)$ terms in all these expansions are neglected, a linear 4×4 system of equations is obtained. By solving this system, we express the third derivative at $i + \frac{1}{2}$ in term of function values at mesh points,

$$h''' = \frac{6}{S_\delta}\left[\frac{h_{i+2}}{\delta_2(\delta_1 + \delta_2)} - \frac{h_{i+1}S_\delta}{\delta_1\delta_2(\delta_0 + \delta_1)} + \frac{h_i S_\delta}{\delta_0\delta_1(\delta_1 + \delta_2)} - \frac{h_{i-1}}{\delta_0(\delta_0 + \delta_1)}\right]. \tag{2.136}$$

Here, $S_\delta = \delta_0 + \delta_1 + \delta_2$. The formula above and a similar expression for the third derivative at $i - 1/2$ have been used in the numerical simulations of fingering instability.

2.5 Notes on Literature

Several well-known reviews address the issue of contact line singularity and approaches to resolving it [20, 55, 70]. The Landau–Levich problem is discussed in detail in the classical book of Levich [88]. Literature on the fingering instability of gravity-driven flow is reviewed by, e.g., Oron et al. [97] and Craster and Matar [40].

Chapter 3
Bubbles and Film Drainage

3.1 Static Interfaces in Channels

Drops and bubbles in macroscale systems are often assumed to be small compared to other relevant length scales describing fluid flow, such as the size of the container. This assumption is in general not justified in microscale applications, e.g., for bubbles transported through long channels of cross-sectional dimensions of $\sim$10–100 μm in a typical microfluidic device. Understanding the effect of geometric confinement on fluid interfaces is essential for modeling such microscale systems as well as a number of other applications including micro heat pipes and fuel cells. A natural first step in this direction is to discuss *static* shapes that gas–liquid interfaces can take in channels of circular and rectangular cross-section.

Consider a gas–liquid interface in a long circular tube as shown in Fig. 3.1. Liquid is partially wetting at the wall, the contact angle is θ. The symmetry suggests that the interface is a two-dimensional surface of revolution, so it is natural to describe its shape in cylindrical coordinates by a function $z = h(r)$, as shown in the sketch in Fig. 3.1. Here all length variables are scaled by the tube radius, R, the z-axis is along the axis of symmetry of the tube.

Based on the general formula for the normal stress balance at a two-dimensional surface discussed in Sect. 1.5* and general arguments used in Sect. 1.2, the static interface shape in our model is described by:

$$\frac{h_{rr}}{(1+h_r^2)^{3/2}} + \frac{h_r}{r(1+h_r^2)^{1/2}} = h\text{Bo} + \text{const.}, \tag{3.1}$$

$$h_r(0) = 0, \quad h_r(1) = \cot\theta. \tag{3.2}$$

Here, we used the formula for the mean curvature of the surface of revolution (1.119), now expressed in nondimensional terms, and $\text{Bo} = \rho g R^2/\sigma$ is the Bond number (gravity vector is in the negative z-direction).

The length scales of cross-sections of long channels discussed in the present chapter (such as R for the circular tube) are assumed to be small enough so that the

V.S. Ajaev, *Interfacial Fluid Mechanics: A Mathematical Modeling Approach*, DOI 10.1007/978-1-4614-1341-7_3, © Springer Science+Business Media, LLC 2012

Fig. 3.1 Sketch of a
gas–liquid interface in a
circular tube; liquid is below
the interface

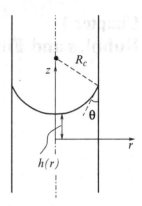

Bond numbers based on these scales are negligible. Then, the condition of interface equilibrium implies simply that the mean curvature is constant. A two-dimensional surface which is part of a sphere clearly satisfies this condition. In order to satisfy the boundary conditions (3.2), the scaled radius of the sphere R_c has to be chosen such that $R_c \cos \theta = 1$ (see Fig. 3.1, the dot denotes the center of the sphere). From simple geometric considerations,

$$h(r) = h_0 + R_c - \sqrt{R_c^2 - r^2}, \qquad R_c = \sec \theta, \tag{3.3}$$

and the constant h_0 is defined by the total volume of liquid in the tube. The same result can of course be obtained by solving the boundary value problem (3.1)–(3.2) with Bo $= 0$. Note that the left-hand side of (3.1) can be expressed as:

$$r^{-1}[rh_r(1 + h_r^2)^{-1/2}]_r. \tag{3.4}$$

For the more general case of nonzero Bond number, the problem defined by (3.1)–(3.2) has to be solved numerically, e.g., using MATLAB.

Manufacturing techniques developed for microscale devices often make it easier to fabricate channels of rectangular rather than circular cross-section. Let us discuss static interface shapes in such channels in the limit of small Bond number, starting with the simple case of a square channel shown in Fig. 3.2. We scale all lengths by the side length of the cross-section and use nondimensional Cartesian coordinates such that x and y are in a horizontal plane and z-axis is in the direction along the axis of the channel. The interface shape is then described by a function $z = h(x, y)$ defined on a square domain of the unit area. The mean curvature H of the interface is given by (1.117), which can be written as:

$$H = \frac{1}{2} \nabla \cdot \hat{\mathbf{n}}, \tag{3.5}$$

Fig. 3.2 Sketch of a
gas–liquid interface in a
square channel

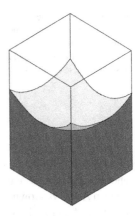

where $\hat{\mathbf{n}} = \hat{\mathbf{n}}(x,y)$ is the x–y projection of the unit normal vector to the interface pointing into the liquid, $\nabla = (\partial/\partial x, \partial/\partial y)$, and all variables are now assumed nondimensional. We note that the vector $\hat{\mathbf{n}}$ can be expressed as $W^{-1}\nabla h$, where $W = (1 + h_x^2 + h_y^2)^{1/2}$.

The conditions of equilibrium are that the mean curvature of the interface is constant, denoted by H in our scaled variables,

$$\nabla \cdot (W^{-1}\nabla h) = 2H, \qquad W = (1 + h_x^2 + h_y^2)^{1/2}, \tag{3.6}$$

everywhere inside a closed domain D and the interface meets the wall at an angle θ, so

$$W^{-1}\nabla h \cdot \mathbf{n} = \cos\theta, \qquad \text{on} \quad \partial D, \tag{3.7}$$

where the boundary of the domain is denoted by ∂D, $\mathbf{n}$ is the unit outward normal vector to the boundary. Based on physical intuition and similarities between this problem and the case of an interface in a circular tube, it may seem obvious that the solution of (3.6)–(3.7) for a square domain D should exist for all values of the contact angle θ. Remarkably, this is not the case, as was shown by Concus and Finn [36]. Based on their result, discussed below, the equilibrium static shape sketched in Fig. 3.2 exists only under the condition $\theta \geq \pi/4$. In general, if channel cross-section is a regular N-sided polygon, then the solution exists only if the contact angle satisfies $\theta \geq \pi/N$. Experiments conducted under the conditions of negligible gravity (at a NASA drop tower) with liquids in channels of hexagonal cross-section verified this prediction [53]: stable equilibrium shapes were observed for $\theta = 48°$ (above the critical value of $\pi/6$), but not for $\theta = 0$ and $\theta = 25°$. In the latter two cases, liquids were observed to rise in the corners up to the top of the container.

Concus and Finn [36] considered the problem (3.6)–(3.7) on a general simply-connected domain D and proved that it does not have a solution if there is a point V on ∂D at which two straight segments of the boundary intersect at an angle 2α such that $\alpha + \theta < \frac{\pi}{2}$. This result can be demonstrated using proof by contradiction. Suppose that a solution exists for $\alpha + \theta < \frac{\pi}{2}$. Following Concus and Finn [36], let us consider a point O on the bisector of the wedge angle and draw a circle of

Fig. 3.3 Sketch of the
geometric construction in the
vicinity of a corner point V
used to prove that (3.6)–(3.7)
does not have solutions if
$\alpha + \theta < \frac{\pi}{2}$

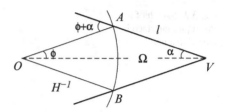

radius H^{-1} centered at O. Assuming O is sufficiently close to the point V, the circle
will intersect with the wedge sides at four points, of which the two closest to V are
denoted by A and B and shown in Fig. 3.3, both at a distance l from V. The arc
between A and B shown in the sketch together with the sides of the wedge form the
boundary of a closed domain Ω. Note that the area of this domain A_Ω can be found
from simple geometric considerations

$$A_\Omega = lH^{-1}\sin(\phi + \alpha) - H^{-2}\phi, \qquad (3.8)$$

where the angle AOV is denoted by ϕ. This angle can be made very small by
reducing l or, equivalently, by changing the location of the point O. So far we did
not impose any restrictions on the location of the point O on the bisector other than
its proximity to V. From now on, we assume that the point O is chosen in such a
way that

$$\phi < \frac{\pi}{2} - \theta - \alpha. \qquad (3.9)$$

Application of the 2D version of divergence theorem (or, equivalently, Green's
theorem in vector form) to (3.6) on the domain Ω results in

$$\int_{\partial\Omega} W^{-1}\nabla h \cdot \mathbf{n}\, ds = 2HA_\Omega. \qquad (3.10)$$

Replacing $W^{-1}\nabla h \cdot \mathbf{n}$ with $\cos\theta$ along the straight line segments, according to the
boundary condition (3.7), and using (3.8), we obtain

$$2l\cos\theta + \int_{\mathscr{C}} W^{-1}\nabla h \cdot \mathbf{n}\, ds = 2l\sin(\phi + \alpha) - 2H^{-1}\phi, \qquad (3.11)$$

where $\mathscr{C}$ denotes the circular arc between A and B. Rearranging terms in the
equation above, we find

$$2l\left[\cos\theta - \sin(\phi + \alpha)\right] = -\int_{\mathscr{C}} W^{-1}\nabla h \cdot \mathbf{n}\, ds - 2H^{-1}\phi. \qquad (3.12)$$

By observing that

$$\left|W^{-1}\nabla h \cdot \mathbf{n}\right| \leq \left|W^{-1}\right|\left|\nabla h\right| = \left(\frac{h_x^2 + h_y^2}{1 + h_x^2 + h_y^2}\right)^{1/2} < 1, \qquad (3.13)$$

Fig. 3.4 Sketch of an
equilibrium bubble shape for
$\theta = 0$ found by Wong
et al. [142] showing half of a
symmetric bubble

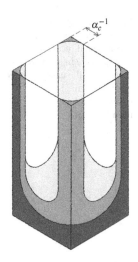

we conclude that the absolute value of the integral on the right-hand side of (3.12)
is less than the length of the curve $\mathscr{C}$ and therefore the right-hand side is negative,
meaning

$$\sin(\phi + \alpha) > \cos \theta \tag{3.14}$$

and therefore

$$\phi + \alpha > \frac{\pi}{2} - \theta. \tag{3.15}$$

This clearly contradicts (3.9). Thus, the problem (3.6)–(3.7) does not have a solution
for $\alpha + \theta < \frac{\pi}{2}$.

Lack of solutions of the boundary value problem (3.6)–(3.7) on a square for
$\theta < \pi/4$ does not rule out the possibility of more complex equilibrium solutions,
such as a bubble shape illustrated in Fig. 3.4 for $\theta = 0$. In contrast to the sketch
shown in Fig. 3.2, liquid is present in the corners in every cross-section of the
channel, so the corners are never completely dry. A technique for solving free
boundary problems describing such shapes has been developed by Wong et al. [142].
They assumed that the macroscopically dry areas on the walls are covered with
ultrathin flat films. In these films, the effects of disjoining pressure discussed in
Sect. 1.7, are important, so the conditions of gas–liquid interface equilibrium have
to be modified to account for them.

Since the mean curvature of the gas–liquid interface is $\frac{1}{2}\nabla \cdot \hat{\mathbf{n}}$, as discussed above,
the interface equilibrium condition incorporating the effects of disjoining pressure
can be written in the form

$$\nabla \cdot \hat{\mathbf{n}} + \Pi\left(\bar{h}\right) = \alpha_c, \tag{3.16}$$

where α_c is a constant, equal to interface curvature in the corner near the middle of
the bubble as shown in the sketch (the interface is nearly two-dimensional there).
The value of α_c is not prescribed in advance but rather found as part of the solution.

Fig. 3.5 Comparison
between the theoretical and
experimental results on
bubble shape in a square
channel for the case of perfect
wetting. From [142],
reprinted with permission
from Elsevier

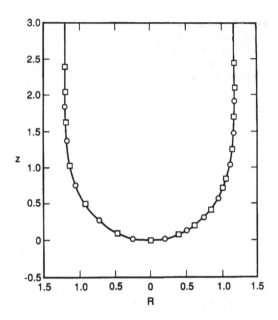

The scaled disjoining pressure Π is a function of $\bar{h}$, the position of the interface measured along the normal to the closest wall of the channel. For the case of perfect wetting, $\Pi(\bar{h})$ is proportional to $\bar{h}^{-3}$, as discussed in Sect. 1.7, although more complicated models of disjoining pressure have been discussed in [142] as well. Away from the walls, the effects of disjoining pressure are negligible, so (3.16) reduces to the condition of constant mean curvature. In the flat ultrathin film regions, curvature is zero, but the condition (3.16) is still satisfied due to the disjoining pressure term, $\Pi(\bar{h})$. This is qualitatively similar to the situation encountered in Sect. 1.7: a flat ultrathin film in equilibrium with a constant-curvature interface.

Wong et al. [142] not only computed complicated static interface shapes using (3.16) for a variety of channels of polygonal cross-sections but also conducted experimental studies of the interface shapes. Figure 3.5 illustrates predictions of the theory (solid line) and experimental results for an air bubble in a square channel (open squares and circles, corresponding to two sets of data obtained from different ends of the bubble). The results correspond to the cross-section of the two-dimensional interface by a plane which passes through two opposite corners of the channel, i.e., makes a 45° angle with the side walls. The channel side length is 0.5 mm and the liquid is perfectly wetting ($\theta = 0$). The comparison with experiments shown in Fig. 3.5 provides a justification for the numerical method developed by Wong et al. [142].

3.2 Bubble Motion in a Circular Tube

Consider a gas bubble in a long circular tube of radius R under the conditions of negligible gravity. As a result of an imposed pressure gradient, liquid moves through the tube, resulting in the motion of the bubble. When the bubble size is large enough, it is observed to take an elongated bullet-type shape sketched in Fig. 3.6. Note that the bubble fills most of the cross-section of the channel, with only a thin film of liquid separating the gas phase and the wall. A mathematical model which describes such bubble shape in the limit of small capillary number has been developed by Bretherton [23]. The capillary number $Ca = \mu U / \sigma$ in this model is based on the velocity of the bubble, U, and pressure is scaled by σ / R.

Assuming the Reynolds number is small, the flow in the liquid around the bubble is described by the Stokes flow equations, as discussed in Sect. 1.5*. Since the flow is axisymmetric, the velocity components are functions of two cylindrical coordinates, r and z, shown in Fig. 3.6, both scaled by the tube radius, R. Away from the walls, the governing equations are reduced to the conditions of capillary statics. For example, consider the equation for the scaled velocity component along the channel, denoted by w,

$$\frac{\partial p}{\partial z} = Ca \nabla^2 w. \tag{3.17}$$

Unless the spatial derivatives on the right-hand side of this equation are large, the small value of Ca implies that $\partial p / \partial z = 0$. The same argument can be applied to the equation for the radial velocity component. Thus, at leading order in Ca the pressure gradient near the bubble surface away from the wall is negligible, so the pressure is constant. These arguments, however, fail when the spatial derivatives of velocity are large, which is likely to happen near the wall of the tube since the dimensional characteristic length there is the thickness of the liquid film separating solid and gas and is much smaller than R. Near the wall, the viscous terms in the equations become important and define the shape of the transition from the capillary statics region to the thin liquid film. The local scales which allow one to describe such transition have been already derived in the context of the Landau–Levich problem in Sect. 2.2. The main complication in the present case is that there are two transition regions, one near the front of the bubble, and the second one at the rear. Let us discuss the front transition region first. The local Cartesian coordinates x and y shown in Fig. 3.6 are scaled by $Ca^{1/3}R$ and $Ca^{2/3}R$, respectively, and the local interface shape is described by the function $y = h(x)$. Note that while the original

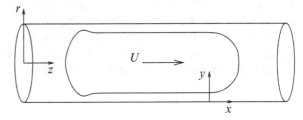

Fig. 3.6 Sketch of a bubble moving in a circular tube showing cylindrical coordinates scaled by the tube radius and local Cartesian coordinates

problem is axisymmetric, the use of local *Cartesian* coordinates is appropriate due to our choice of length scales. The coordinate system is moving with the bubble, so the interface profile is independent of time. Using the same arguments as in Sect. 1.4, the general flow equations are now reduced to the lubrication-type system of the form

$$p_x = u_{yy},$$ (3.18)

$$p_y = 0,$$ (3.19)

$$u_x + v_y = 0.$$ (3.20)

Since p is clearly not a function of y, the first equation can be integrated twice in y, resulting in

$$u = \frac{p_x}{2}(y^2 - 2yh) - 1.$$ (3.21)

Here, we used the condition of no-slip at the solid (which in our moving reference frame implies $u(x,0) = -1$). The capillary pressure jump at the interface results in the boundary condition of the same form as used in Sect. 1.4:

$$p - p_g = -h'' \quad \text{at} \quad y = h(x),$$ (3.22)

where p_g is the pressure in the gas phase inside the bubble, primes are used to denote derivatives with respect to x. Integrating (3.20) in y, we obtain

$$\int_0^h u \, dy = c_1,$$ (3.23)

where c_1 is a constant. Substituting (3.21) and (3.22) into this equation and integrating the result, we find

$$h^3 h''' = 3(h - c_1).$$ (3.24)

Since in the flat film region all derivatives of h are zero, the constant c_1 is equal to the scaled thickness of this film, b, which has to be determined as part of the solution. Then, the equation for $h(x)$ becomes

$$h^3 h''' = 3(h - b).$$ (3.25)

Note that aside from differences in notation this equation is identical to the one determining the local interface shape for the Landau–Levich problem, (2.92), derived in Sect. 2.2. However, the boundary condition at ∞ is different since the interface curvature has to match the zero-Bond-number capillary statics solution obtained in Sect. 3.1 rather than the solution from Sect. 1.2. Using the equation for the radius of curvature from (3.3) with $\theta = 0$ (as the only value consistent with our scaling in the limit of Ca $\to 0$), we obtain the matching condition

$$h''(\infty) = 1.$$ (3.26)

Fig. 3.7 Interface shapes in the transition regions near the front (*top graph*) and the rear end (*bottom graph*) of the bubble moving through a circular tube

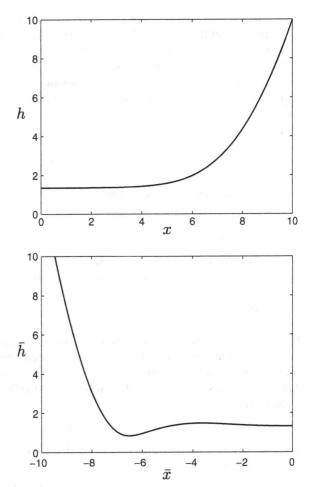

The boundary conditions in the flat film are in the form,

$$h(-\infty) = b, \quad h'(-\infty) = h''(-\infty) = 0. \tag{3.27}$$

The problem defined by (3.25) and (3.27) was solved using the numerical method described in detail in Sect. 2.2 (with the corresponding MATLAB code in Sect. B.3). The film thickness b is then found based on the matching condition, (3.26), resulting in the value

$$b = 1.3375. \tag{3.28}$$

Typical local interface shape for the front transition region, $y = h(x)$, is shown in Fig. 3.7 (top). In dimensional terms, our result implies that when a long bubble moves in a circular tube at a speed U, it is separated from the wall of the tube by a thin liquid film of thickness $1.3375R\mathrm{Ca}^{2/3}$. Note that this expression is obtained by considering only the front part of the bubble.

To find the interface shape in the rear transition region of the bubble, consider local Cartesian coordinates there, denoted by $\bar{x}$ and $\bar{y}$, scaled by $\mathrm{Ca}^{1/3}R$ and $\mathrm{Ca}^{2/3}R$, respectively ($\bar{x}$ is in the same direction as x). Following the same steps as in the above derivation for the front transition region, we arrive at the equation for the local interface shape $\bar{y} = \bar{h}(\bar{x})$,

$$\bar{h}^3 \bar{h}''' = 3(\bar{h} - b). \tag{3.29}$$

It is convenient to introduce new variables $\hat{h} = \bar{h}/b$ and $\hat{x} = -3^{1/3}\bar{x}/b$, so that (3.29) takes the form

$$\hat{h}''' = \frac{1 - \hat{h}}{\hat{h}^3}, \tag{3.30}$$

where primes now denote derivatives with respect to $\hat{x}$. Using the same variables, the conditions in the flat film region and far away from the wall become

$$\hat{h}(-\infty) = 1, \quad \hat{h}'(-\infty) = \hat{h}''(-\infty) = 0, \quad \hat{h}''(\infty) = b/3^{2/3}. \tag{3.31}$$

Note that since there are four conditions and the value of b is prescribed, the existence of the solution at this point is an assumption which is justified by constructing the solution numerically. If the origin of coordinates is chosen such that $\zeta = \hat{h} - 1$ is small near $\hat{x} = 0$, then ζ satisfies the linearized version of (3.30), in the form

$$\zeta''' = -\zeta. \tag{3.32}$$

Solutions of this equation which decay at $\hat{x} \to -\infty$ are written as:

$$\zeta = \alpha_1 e^{\frac{1}{2}\hat{x}} \cos \frac{\sqrt{3}}{2}\hat{x} + \alpha_2 e^{\frac{1}{2}\hat{x}} \sin \frac{\sqrt{3}}{2}\hat{x}, \tag{3.33}$$

where α_1 and α_2 are constants. One of them can be taken as an arbitrary small number ($\alpha_1 = 10^{-3}$ in our simulations), which is equivalent to fixing the location of the origin of coordinates. The second constant is then found by the standard shooting method with the condition of known scaled curvature, equal to $b/3^{2/3}$, at the right endpoint of the computational domain. The boundary conditions at the left endpoint, $\hat{x} = 0$, are formulated in terms of α_1 and α_2 by evaluating the function $\hat{h} = 1 + \zeta$ and its derivatives at $\hat{x} = 0$. The resulting interface shape, represented in terms of the coordinates $\bar{x}$ and $\bar{y}$, is shown in Fig. 3.7 (bottom). Comparison between the shapes of the local solutions in the top and bottom parts of Fig. 3.7 explains the characteristic bullet-type shape of the bubble sketched in Fig. 3.6. Note that the bubble shapes for small Ca are in agreement with experimentally observed ones.

3.3 Bubbles in Square Tubes

A bubble moving in a circular tube discussed in the previous section fills the entire cross-section of the tube, with only thin film (of thickness proportional to $Ca^{2/3}$) separating the gas phase and the solid. However, when the models of this type are compared with experimental data on bubble motion in *square* tubes, significant discrepancies can be observed. The origin of these discrepancies has been investigated by Wong et al. [143, 144]. Qualitatively, when a bubble moves in a square channel, liquid can flow around the bubble through corners, which means that the resistance to the bubble motion is significantly less than for the case of bubble in a circular tube. Wong et al. [143] developed an asymptotic theory describing bubbles moving in square and rectangular channels. Bubble shape is dominated by capillary forces away from the wall in their model, so it is natural to assume that the interface shape there will be the same as the static shape discussed in Sect. 3.1 and sketched in Fig. 3.4. However, in the case of a moving bubble, sketched in Fig. 3.8, there are liquid films on the walls formed by the same liquid entrainment mechanism as in the Bretherton problem from the previous section.

The scales used to describe liquid entrainment near the leading edge of the bubble are the same as in the previous section, so the thickness of the entrained film there scales as $Ca^{2/3}$. However, a different type of transition is observed between the meniscus and a thin film near the sides of the bubble away from the leading edge. Capillary pressure in this region creates a draining flow from the film to the meniscus in the corner (assuming that the film is macroscopic, so the effects of disjoining pressure are not significant enough to compensate for this flow). This results in complicated rearrangement of the film; its characteristic thickness scales as $Ca^{4/3}$.

Wong et al. [144] also investigated a relationship between the pressure gradient along the channel and the velocity of bubble motion. The key to obtaining such relationship is a model of flow in the corners through which the liquid can bypass the bubble. A parallel flow approximation was used in the numerical solution in [144],

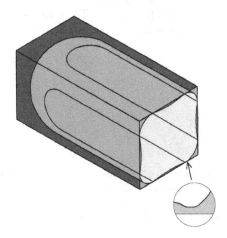

Fig. 3.8 Sketch of a bubble moving in a square channel. From [5]

resulting in a Poisson's equation for the velocity in the direction along the channel axis. The simplified numerical model of [144] was later shown to be in agreement with more recent numerical simulations of confined bubbles, as discussed in more detail, e.g., in [5].

3.4   Bubble Pressed Against a Solid Wall

In practical applications of drops and bubbles on microscale their shapes can be time dependent rather than steady. Consider for example an experimental set-up in which bubble shape is slowly evolving due to interaction with a solid wall, as sketched in Fig. 3.9. A bubble is maintained at a constant pressure at the tip of a circular tube and the distance between the tube and the solid wall is gradually reduced, at a speed V. As a result, the bubble is deformed by the wall and viscous flow is generated in the liquid gap between the bubble and the solid surface. Suppose U is a characteristic velocity of this flow, R_0 is the bubble radius when it is far from the wall and its shape can be described as part of a sphere (assuming the Bond number is zero). If both the capillary number, $Ca = \mu U / \sigma$, and the liquid gap thickness scaled by R_0, are small, it is natural to describe the flow in the liquid using a lubrication-type approach. However, if we follow the approach of Sect. 1.4 and simply rescale the nondimensional gap thickness as a certain power of Ca while using R_0 for the radial scale, it will not be possible to match our solution to the interface away from the wall where the bubble shape is dominated by capillary forces. This issue is essentially the same as the one encountered in the context of the Landau–Levich problem in Sect. 2.2, so it can be resolved by using the set of scales introduced there. They correspond to a bubble deformed locally in the region of a characteristic size $Ca^{1/3}R_0$ in the radial direction and $Ca^{2/3}R_0$ in the vertical direction. This is where viscous forces are of the same order as the surface tension forces; outside of this region the shape of the bubble is dominated by capillarity.

We use local cylindrical coordinates shown in Fig. 3.9, with z scaled by $Ca^{2/3}R_0$ and r scaled by $Ca^{1/3}R_0$. The velocity components u (in the radial direction) and w (in the vertical direction) are scaled by U and $Ca^{1/3}U$, respectively. The standard

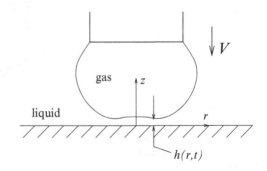

Fig. 3.9 Sketch of a bubble pressed against a solid wall. Nondimensional cylindrical coordinates r and z are shown

system of Navier–Stokes and continuity equations in cylindrical coordinates (see, e.g., Appendix A.6 in [1]) can be simplified by using the assumption of axial symmetry and neglecting all terms which are small in the limit of Ca $\rightarrow$ 0, resulting in an axisymmetric version of the standard lubrication-type equations,

$$-p_r + u_{zz} = 0, \tag{3.34}$$

$$-p_z = 0, \tag{3.35}$$

$$r^{-1}(ru)_r + w_z = 0. \tag{3.36}$$

The effects of gravity are neglected, assuming that the Bond number is small; p is the difference between pressure in the liquid and inside the bubble, scaled by σ/R_0.

The nondimensional thickness of the liquid film between the bubble and the solid wall is described by a function $z = h(r,t)$, where time t is scaled by $\mathrm{Ca}^{-1/3}d/U$. The interfacial stress conditions are

$$p = h_{rr} + r^{-1}h_r \quad \text{at} \quad z = h(r,t), \tag{3.37}$$

$$u_z = 0 \quad \text{at} \quad z = h(r,t). \tag{3.38}$$

The expression on the right-hand side of the formula for pressure in (3.37) is obtained from the general formula for the mean curvature of a surface of revolution, (1.119), by writing it in nondimensional form and neglecting all terms which are small in the limit of Ca $\rightarrow$ 0.

Kinematic boundary condition in cylindrical coordinates is in the form

$$h_t + uh_r - w = 0 \quad \text{at} \quad z = h(r,t) \tag{3.39}$$

and the usual boundary conditions at the solid surface (assuming negligible slip length) are expressed by

$$u = w = 0 \quad \text{at} \quad z = 0. \tag{3.40}$$

Following the same steps as in Sect. 1.4, the lubrication-type system of equations and boundary conditions formulated here leads to an evolution equation for the film thickness, $h(r,t)$, in the form

$$h_t + (3r)^{-1}\left[rh^3\left(h_{rr} + r^{-1}h_r\right)_r\right]_r = 0. \tag{3.41}$$

This equation is considered on the interval $[0, L]$ with L chosen large enough so that the boundary conditions at $r = L$ can be imposed based on the expected behavior of the solution at $r \rightarrow \infty$, in the form

$$h_t(L,t) = -\hat{V}(t), \quad h_{rr}(L,t) + L^{-1}h_r(L,t) = 1. \tag{3.42}$$

Here, the first condition states that the interface moves down with a prescribed nondimensional velocity, equal to the downward velocity of the tube scaled by $Ca^{1/3}U$. In the simulations below we model an experimental procedure in which the tube moves a fixed distance downward and then stops, by choosing

$$\hat{V}(t) = \begin{cases} 1, & 0 \leq t \leq \Delta t, \\ 0, & t > \Delta t. \end{cases} \tag{3.43}$$

The second condition in (3.42) is that of matching our solution to the capillary static interface shape, which is part of a sphere of nondimensional radius 1. Conditions of axial symmetry are specified at $r = 0$:

$$h_r(0,t) = 0, \qquad h_{rrr}(0,t) = 0. \tag{3.44}$$

The initial condition is in the form

$$h(r,0) = h_0 + \frac{1}{4}r^2. \tag{3.45}$$

A numerical finite-difference procedure similar to the one discussed in Sect. 1.6 can be used to solve the system defined by (3.41) and conditions (3.42)–(3.45). The only significant difference is that the point $r = 0$ requires special treatment. The second term on the left-hand side of (3.41) cannot be evaluated directly at $r = 0$ because of the r^{-1} factor. To resolve this difficulty, we consider a small but finite value of r and use the following Taylor expansions:

$$h_{rr}(r,t) = h_{rr}(0,t) + \frac{1}{2}h_{rrrr}(0,t)r^2 + O(r^3), \tag{3.46}$$

$$h_r(r,t) = h_{rr}(0,t)r + \frac{1}{6}h_{rrrr}(0,t)r^3 + O(r^4). \tag{3.47}$$

Here we used (3.44) to eliminate some of the terms of the full Taylor series. $O(r^m)$ $(m = 3,4)$ denotes a function $f(r,t)$ such that

$$\lim_{r \to 0} \frac{f(r,t)}{r^m} \tag{3.48}$$

exists. Substituting the expansions for the derivatives into (3.41), we find

$$h_t(r,t) + \frac{8}{9}[h(0,t)]^3 h_{rrrr}(0,t) + O(r) = 0 \quad \text{as} \quad r \to 0. \tag{3.49}$$

This formula is used to evaluate the derivative h_t at the left endpoint of the computational domain.

Typical numerical results for $h(r,t)$ are shown in Fig. 3.10 (for $h_0 = 1$, $L = 6$) at three different times specified in the legend. The initially constant-curvature

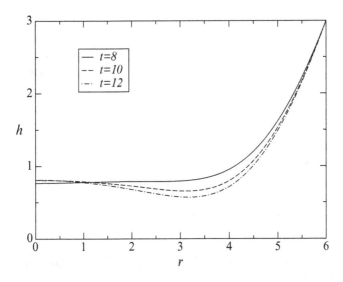

Fig. 3.10 Snapshots of the interface showing dimple formation, obtained from the numerical solution of (3.41)

interface, defined by (3.45), first flattens as it approaches the wall and then deforms to develop a so-called dimple. The dimple shape provides a capillary pressure gradient, from higher pressure at the axis of symmetry to lower pressure near the point of local minimum, needed for continuous drainage of the liquid film.

If the numerical simulation is run for much longer times than shown in Fig 3.10, the liquid film will drain and the minimum value of h will gradually approach zero. In experiments, however, a stable film of finite thickness can form due to effects of disjoining pressure, as discussed in Sect. 1.7. Our model contradicts these experiments because it does not incorporate the effects of disjoining pressure. In order to account for them in our framework, it is sufficient to modify the normal stress boundary condition, (3.37), by adding a disjoining pressure term:

$$p = h_{rr} + r^{-1}h_r + \alpha h^{-3} \quad \text{at} \quad z = h(r,t), \qquad (3.50)$$

where $\alpha = -A/\sigma Ca^2 R_0^2$ and A is the Hamaker constant introduced in Sect. 1.7 and assumed negative here.

Then, the modified evolution equation is

$$h_t + (3r)^{-1} \left[rh^3 \left(h_{rr} + r^{-1}h_r + \alpha h^{-3} \right)_r \right]_r = 0. \qquad (3.51)$$

Numerical solution of this equation with the conditions (3.42)–(3.45) and $\alpha = 10^{-3}$ shows initial behavior very similar to the one shown in Fig. 3.10: flattening and formation of a dimple. However, at longer times ($t \sim 10^2$), the dimple gradually disappears and the film flattens, as shown in Fig. 3.11. Eventually the solution

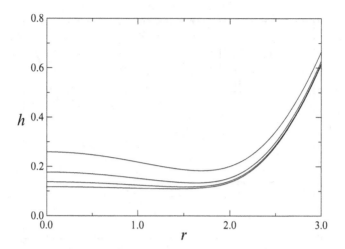

Fig. 3.11 Solutions of (3.51) with $\alpha = 10^{-3}$ at large times showing flattening of the interface due to disjoining pressure

approaches an equilibrium state such that there is a flat film of scaled thickness $\sim \alpha^{1/3}$ (in which disjoining pressure is important) in equilibrium with a meniscus of constant-curvature and the total pressure given by (3.50) is constant everywhere. Equating the pressure in the flat film (zero curvature) and the capillary pressure jump away from the interface (constant-curvature, negligible disjoining pressure) leads to the equilibrium condition

$$\frac{\alpha}{h^3} = 2. \tag{3.52}$$

Alternatively, this condition can be written in dimensional terms as:

$$-\frac{A}{h^{*3}} = \frac{2\sigma}{R_0}. \tag{3.53}$$

By measuring the thickness of the ultrathin film h^*, e.g., using the methods discussed in Sect. 1.7, and the radius of the interface away from the walls, R_0, the value of the Hamaker constant can then be found, as was done in the pioneering works of Derjaguin et al. [44, 45]. It is important to note that the condition (3.53) is based on the assumption that the flat ultrathin film is formed due to London–van der Waals dispersion forces. In experiments, other physical effects, e.g., due to the presence of electric charges, as discussed below in Chap. 4, can affect the equilibrium conditions.

3.5 Rupture in Liquid Films

3.5.1 Unstable Ultra Thin Films

The discussion of film drainage in Sect. 3.4 was based on the assumption that when the film becomes thin enough for the disjoining pressure to be important, it is stabilized due to the negative value of the Hamaker constant. In general, the Hamaker constant A (with absolute values typically on the order of 10^{-21} J) depends on the properties of *both* liquid and solid. It also has a weak (and often neglected) dependence on the properties of the gas phase. The equilibrium configuration seen in our simulations of film drainage in Fig. 3.11 is possible only when the Hamaker constant is negative (and thus according to (1.129) Π is positive) so that the disjoining pressure in the flat film balances the capillary pressure jump across the liquid–gas interface away from the solid surface. Theoretical models discussed in Sect. 1.7 show that the Hamaker constant can also be positive. This is confirmed by experimental studies, e.g., for polystyrene films on silicone oxide. Experimental studies of film drainage in this case indicate that the films become unstable when they are thin enough for the disjoining pressure to be important: small perturbations of the interface shape grow and eventually result in rupture of the film. Despite the fact that relatively few liquid–solid combinations have positive Hamaker constant, we discuss this case in detail since it allows us to illustrate a number of important mathematical techniques which are applicable to a wide range of problems, including instabilities arising in more complicated models of disjoining pressure. The process of the development of instability can be described in the framework of a lubrication-type model [27, 138] similar to the one discussed in Sect. 1.4. Indeed, due to very small value of film thickness it is natural to assume that the ratio of the thickness to typical horizontal length scales is small. We consider a thin liquid film on a flat horizontal solid surface and use the system of lubrication equations (1.46)–(1.48) as the starting point for developing our model (as in Sect. 1.4, the model here is two-dimensional, with the x-axis along the substrate and the y-axis normal to it). However, the boundary conditions are different from the ones used in Sect. 1.4: the normal stress condition is modified by the presence of disjoining pressure and the thermocapillary contribution to the tangential stress is set to zero since there are no temperature gradients:

$$p - p_0 = -\sigma \frac{\partial^2 h}{\partial x^2} + \frac{A}{h^3} \quad \text{at} \quad y = h(x,t), \tag{3.54}$$

$$\frac{\partial u}{\partial y} = 0 \quad \text{at} \quad y = h(x,t). \tag{3.55}$$

To make a better connection with the derivation from Sect. 1.4, we start with *dimensional* variables here and then use tildas later to denote their scaled versions. A procedure similar to the derivation of (1.66) can be used to obtain the following dimensional integral mass balance

$$\frac{\partial h}{\partial t} + \frac{\partial}{\partial x}\left(\int_0^h u\,dy\right) = 0. \tag{3.56}$$

By integrating (1.46) with the no-slip condition ($u = 0$) at the solid boundary and the condition (3.55) at the gas–liquid interface we find the following dimensional horizontal velocity profile,

$$u = \frac{1}{2\mu}\frac{\partial p}{\partial x}\left(y^2 - 2hy\right). \tag{3.57}$$

Substitution of this profile into (3.56) leads to

$$\frac{\partial h}{\partial t} - \frac{1}{3\mu}\frac{\partial}{\partial x}\left(h^3\frac{\partial p}{\partial x}\right) = 0. \tag{3.58}$$

Assuming that the effects of gravity are negligible and using (3.54) to express pressure in the liquid in terms of the interface shape, $h(x,t)$, we obtain

$$\frac{\partial h}{\partial t} + \frac{1}{3\mu}\frac{\partial}{\partial x}\left(\sigma h^3\frac{\partial^3 h}{\partial x^3} + \frac{3A}{h}\frac{\partial h}{\partial x}\right) = 0. \tag{3.59}$$

In contrast to the model of film rupture driven by thermocapillarity from Sect. 1.4, the present model has time-independent solutions corresponding to uniform film thickness and quiescent liquid. In other words, $h(x,t) = d$ is a solution of (3.59) for any constant d as long as the boundary conditions (if a film of finite extent is considered) are consistent with this solution and the initial condition is chosen to be $h(x,0) = d$. However, uniform films may still be impossible to observe in experiments (except for a very short time) if small random perturbations in film thickness have a tendency to grow. Since small perturbations are always present in experiments, the uniform film solution will be rapidly destroyed when such perturbations grow in time. To make predictions about evolution of a small perturbation of the interface shape we consider solutions of (3.59) in the form

$$h = d + \zeta(x,t). \tag{3.60}$$

Since we are interested in small perturbations, $|\zeta| \ll d$, their time evolution can be described by a linearized version of (3.59), obtained by neglecting all terms which are nonlinear in ζ and its derivatives:

$$\frac{\partial \zeta}{\partial t} + \frac{\sigma d^3}{3\mu}\frac{\partial^4 \zeta}{\partial x^4} + \frac{A}{\mu d}\frac{\partial^2 \zeta}{\partial x^2} = 0. \tag{3.61}$$

A perturbation ζ can be expressed as a superposition of sinusoidal functions (by using Fourier series expansion in x for a finite size domain or the Fourier transform in the same variable for an infinite domain). Since (3.61) is linear, its solution will be a superposition of solutions corresponding to each sinusoidal perturbation.

Therefore, conclusions about stability of the solution can be made by analyzing time evolution of a sinusoidal perturbation as a function of its wavenumber k. If the initial shape of the perturbation is

$$\zeta(x,0) = \hat{\zeta}_0 \cos kx, \tag{3.62}$$

its evolution is described by $\zeta = \hat{\zeta}(t) \cos kx$. By substituting this into (3.61), we obtain the following initial value problem for $\hat{\zeta}(t)$:

$$\frac{d\hat{\zeta}}{dt} = \left(\frac{A}{\mu d} k^2 - \frac{\sigma d^3}{3\mu} k^4 \right) \hat{\zeta}, \quad \hat{\zeta}(0) = \hat{\zeta}_0. \tag{3.63}$$

The solution of this initial value problem is $\hat{\zeta}(t) = \hat{\zeta}_0 e^{\gamma t}$, where the growth rate γ is defined by:

$$\gamma = \frac{A}{\mu d} k^2 - \frac{\sigma d^3}{3\mu} k^4. \tag{3.64}$$

According to this formula, only perturbations with wavenumbers satisfying $k^2 < 3A/\sigma d^4$ grow in time. Equation (3.64) can be written in a convenient nondimensional form as

$$\tilde{\gamma} = \tilde{k}^2 - \tilde{k}^4, \tag{3.65}$$

where

$$\tilde{\gamma} = \gamma \frac{\sigma \mu d^5}{3A^2}, \quad \tilde{k}^2 = k^2 \frac{\sigma d^4}{3A}. \tag{3.66}$$

The nondimensional growth rate $\tilde{\gamma}$ is plotted as a function of $\tilde{k}$ in Fig. 3.12. Note that the film is always unstable for $A > 0$, but the growth rate of the instability can be very small for large d. The fastest growth corresponds to the point of maximum of this curve, $\tilde{k}^2 = \frac{1}{2}$, which in dimensional terms can be expressed as $k^2 = 3A/(2\sigma d^4)$. The fastest growing mode of the instability usually defines the characteristic length scale of the perturbations observed in experiments.

In general, the results of linear stability analysis are typically expressed in the form of an equation which relates the growth rate of an instability to its wavenumber, often called a dispersion relation. Equations (3.64) and (3.65) are examples of dispersion relations in dimensional and nondimensional form, respectively.

3.5.2 Nondimensional Evolution Equation

The linearized equation (3.61) describes the initial stage of instability development, but to predict significant departures from the uniform thickness solution one has to use the original nonlinear evolution equation, (3.59). Let us rewrite the latter in nondimensional form. Since the fastest growing perturbation determines the typical

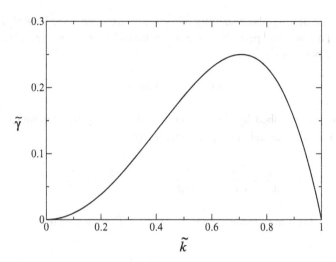

Fig. 3.12 Growth rate of a sinusoidal perturbation as a function of its wavenumber in nondimensional form

horizontal length and evolution time of deformations observed in experiments, the characteristic horizontal length scale and time scale of the nondimensional formulation should be of the same order; we define nondimensional variables by:

$$\tilde{x} = \frac{x}{d^2}\sqrt{\frac{3A}{\sigma}}, \quad \tilde{y} = \frac{y}{d}, \quad \tilde{t} = t\frac{3A^2}{\sigma\mu d^5}. \tag{3.67}$$

The ratio of the length scales in the horizontal and vertical directions is equal to d^*/d, where $d^* = (3A/\sigma)^{1/2}$ is typically less than 1 nm. Thus, even for very thin liquid films the ratio of these scales is small.

The nondimensional version of (3.59) with the scaled variables defined by (3.67) is written as:

$$\frac{\partial \tilde{h}}{\partial \tilde{t}} + \frac{\partial}{\partial \tilde{x}}\left(\tilde{h}^3\frac{\partial^3 \tilde{h}}{\partial \tilde{x}^3} + \frac{1}{\tilde{h}}\frac{\partial \tilde{h}}{\partial \tilde{x}}\right) = 0. \tag{3.68}$$

Let us now discuss the choice of the domain on which this equation will be solved numerically. In experiments, the extent of the film is usually much larger than the wavelength of the fastest growing perturbation. The latter is equal to $\lambda = 2\pi\sqrt{2}$ in nondimensional terms (since the maximum of $\tilde{\gamma}(\tilde{k})$ in (3.65) corresponds to $\tilde{k} = \frac{1}{\sqrt{2}}$). Thus, it may seem that the computational domain has to be much larger than $2\pi\sqrt{2}$. However, since the growth of the perturbations is dominated by the fastest growing mode, it turns out that the rupture process can be described reasonably well by simulations conducted on any domain large enough to fit the fastest growing instability mode. For example, by obtaining the numerical solution on the domain $[-\pi\sqrt{2}, \pi\sqrt{2}]$ and then extending it periodically over the entire real

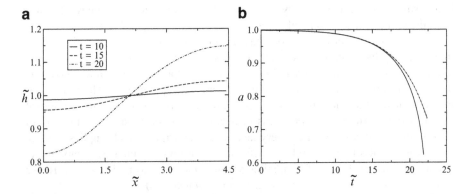

Fig. 3.13 (**a**) Snapshots of the interface shape at different times obtained by solving (3.68) numerically; (**b**) Initial evolution of scaled minimum thickness of the film found from the numerical solution of (3.68) (*solid line*) and predicted by the linear theory (*dot–dashed line*)

axis, one can get not only qualitative but also quantitative insight into the behavior of the interface on a much larger domain. In the simulations of this section, we only consider solutions which are symmetric with respect to the origin, so that it is sufficient to solve (3.68) on the domain $[0, \pi\sqrt{2}]$ with the boundary conditions

$$\frac{\partial \tilde{h}}{\partial \tilde{x}}(0, \tilde{t}) = \frac{\partial^3 \tilde{h}}{\partial \tilde{x}^3}(0, \tilde{t}) = \frac{\partial \tilde{h}}{\partial \tilde{x}}\left(\pi\sqrt{2}, \tilde{t}\right) = \frac{\partial^3 \tilde{h}}{\partial \tilde{x}^3}\left(\pi\sqrt{2}, \tilde{t}\right) = 0. \tag{3.69}$$

The initial condition is in the form

$$\tilde{h}(\tilde{x}, 0) = 1 - \zeta_0 \cos\left(\frac{\tilde{x}}{\sqrt{2}}\right). \tag{3.70}$$

A straightforward modification of the code from Sect. 1.6 allows us to obtain snapshots of the perturbed interface, $\tilde{h}(\tilde{x}, \tilde{t})$, shown in Fig. 3.13a. The amplitude of the initial sinusoidal perturbation in this simulation is $\zeta_0 = 10^{-3}$. The interface evolution clearly speeds up with time and departures from the sinusoidal shape are seen at later times: the dot–dashed curve departs from the average value of $\tilde{h} = 1$ significantly more near the origin than near the right endpoint of the domain.

With our choice of the initial condition, the minimum nondimensional film thickness, denoted by a, is always reached at the point $\tilde{x} = 0$. The linear theory discussed in the previous subsection predicts that a will change with time according to

$$a = 1 - \zeta_0 e^{\tilde{t}/4}. \tag{3.71}$$

The result from the numerical solution of (3.68) at the initial stages of evolution compares very well with this prediction, as illustrated in Fig. 3.13b. However, at later times the departures of the interface shape from the average uniform film are no longer small and the linear theory prediction (dot–dashed line in the figure) deviates from the numerical solution (solid line).

The standard numerical methods based on uniform finite-difference discretization in space, discussed in detail in Sect. 1.6, rapidly lose accuracy when the value of a becomes small. Thus, they fail to accurately compute the total time it takes for the film to rupture, an important parameter for practical applications. The reasons for the numerical difficulties are twofold, as can be seen from analyzing the numerical data just before the catastrophic loss of accuracy occurs. First, the values of all even spatial derivatives of $\tilde{h}$ are very large near the origin, which implies that the numerical method is likely to have difficulties resolving the interface shape there. Second, the rate of change of a with time becomes increasingly higher as a decreases (this tendency can be seen even in Fig. 3.13b for a near 0.6).

The problem with the spatial discretization is localized: the derivatives are very high only near the origin. As a remedy for this, it is natural to use a nonuniform spatial mesh which has more points in the region where the function has high derivatives. The rapid variation of a with time can be resolved if progressively smaller time steps are used for smaller a. However, there are many different ways to introduce nonuniform meshes and variable time stepping and not all of them are equally efficient. In order to find a suitable approach for our equation, it is useful to first investigate the structure of the solution for small a.

3.5.3 Similarity Solution Near the Point of Rupture

In this subsection, we illustrate how one can gain useful insights into the behavior of solutions of (3.68), e.g., the dependence of the scaled minimum thickness a on time for small a, without doing any numerical computations. To accomplish this goal, we follow Zhang and Lister [147] and consider solutions of a special type

$$\tilde{h}(\tilde{x}, \tilde{t}) = m(\tilde{t})H(\eta), \qquad (3.72)$$

often called *similarity* (or *self-similar*) solutions. Here, $\eta = \tilde{x}/l(\tilde{t})$ is called the similarity variable; in order to define the yet unknown functions $m(\tilde{t})$ and $l(\tilde{t})$ we assume that the solution for small a is independent from the initial and boundary conditions (which is essentially assuming that the local behavior for small a is not significantly affected by the solution away from the point where the film eventually ruptures). The similarity solution discussed here is only valid at small a, not over the entire time domain (the initial evolution, e.g., shown in Fig. 3.13, does depend on the initial and boundary conditions). Suppose that for small a the minimum thickness behaves as:

$$a(\tilde{t}) = C_1 \left(\tilde{t}_R - \tilde{t}\right)^\alpha . \qquad (3.73)$$

Here, we introduce the scaled rupture time, $\tilde{t}_R$, at which the value of a reaches zero; C_1 and α are constants. Based on (3.67), the dimensional version of (3.73) is

$$a^* = d \left(\frac{3A^2}{\sigma \mu d^5} \right)^{\alpha} C_1 (t_R - t)^{\alpha}, \tag{3.74}$$

where a^* and t_R are the dimensional values of the minimum thickness of the film and the rupture time. According to our assumption, $a^*(t)$ should be independent of the initial conditions and therefore independent of d. This is possible only when $\alpha = 1/5$, so we conclude that

$$a(\tilde{t}) \sim (\tilde{t}_R - \tilde{t})^{1/5}. \tag{3.75}$$

This behavior follows from (3.72) if we choose $m(\tilde{t}) = (\tilde{t}_R - \tilde{t})^{1/5}$. To determine $l(\tilde{t})$, we assume

$$l(\tilde{t}) = (\tilde{t}_R - \tilde{t})^{\beta} \tag{3.76}$$

with a constant β. Since the dimensional version of (3.76) should be independent of d, we conclude that $\beta = 2/5$ and thus

$$l(\tilde{t}) = (\tilde{t}_R - \tilde{t})^{2/5}. \tag{3.77}$$

An important benefit of the similarity solution (3.72) is that it can be used to find a local time-dependent interface shape near the point of rupture by solving an ordinary differential equation for $H(\eta)$ instead of the partial differential equation for $\tilde{h}(\tilde{x}, \tilde{t})$. This ordinary differential equation can be obtained by substituting the similarity ansatz, (3.72), into (3.68). However, since we are interested in finding numerical solutions over the entire time domain and not just near the rupture point, we do not pursue this approach any further. A detailed discussion of the equation for $H(\eta)$ can be found in Zhang and Lister [147].

Equations (3.75) and (3.77) for $a(\tilde{t})$ and $l(\tilde{t})$ suggest an approach to the numerical solution of the original equation (3.68) which is suitable for accurately and efficiently resolving the behavior of the solution near the point of rupture. According to (3.75), the relevant timescale of change of the solution near the point of rupture is proportional to a^5, so to resolve this scale we need the time steps which decrease accordingly; in the simulation below we choose variable time step $\Delta \tilde{t} = 0.01 a^5$. Comparing (3.75) and (3.77), we conclude that the horizontal length scale near the point of rupture decreases as a^2, so the mesh size near the point of rupture $\Delta \tilde{x}$ should decrease as a^2 for small a. However, there is no need to use very fine mesh away from the point of rupture, so we introduce a nonuniform mesh defined by:

$$x_i = a^2 \sinh(s_i), \quad s_i = \frac{i-1}{N} \sinh^{-1} \left(\frac{\pi \sqrt{2}}{a^2} \right), \quad i = 1, \ldots, N+1. \tag{3.78}$$

The standard finite difference approximations of derivatives are no longer applicable when a nonuniform mesh is used, but their generalization is straightforward, as

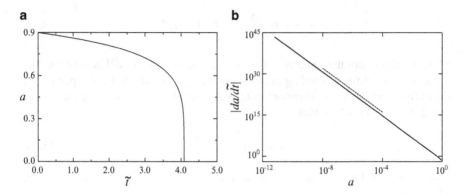

Fig. 3.14 (a) Minimum film thickness as a function of time in scaled coordinates; (b) A plot illustrating self-similar behavior near the point of rupture

discussed in detail in Sect. 2.4.5. Here we just use the results from that section for the particular mesh defined above. The locations of the mesh points change at every time step; the new values of the function $\tilde{h}$ are found by second-order interpolation

$$h_i^{\text{new}} \approx h_i + \tilde{h}_{\tilde{x}}(x_i,t)(x_i^{\text{new}} - x_i) + \frac{1}{2}\tilde{h}_{\tilde{x}\tilde{x}}(x_i,t)(x_i^{\text{new}} - x_i)^2, \tag{3.79}$$

where the spatial derivatives at the point x_i are evaluated using the formulas from Sect. 2.4.5. The method described here is a slightly modified version of the approach developed by Zhang and Lister [147].

Numerical results obtained by solving (3.68) with the initial condition given by (3.70), $\tilde{\zeta}_0 = 0.1$, and $N = 100$, are presented in Fig 3.14. Part "a" of this figure shows the minimum thickness as a function of time in scaled coordinates. Clearly, the rate of change of a increases sharply prior to rupture. To verify (3.75), we note that it implies

$$\left|\frac{da}{d\tilde{t}}\right| \sim (\tilde{t}_R - \tilde{t})^{-4/5} \sim a^{-4} \tag{3.80}$$

so the plot of $|da/d\tilde{t}|$ versus a in log–log coordinates should be a straight line with the slope of -4. This is indeed observed in Fig 3.14b over at least ten orders of magnitude (the dashed line has a slope of -4 and is shown for comparison). Thus, the behavior predicted by the similarity solution is indeed observed in the numerical simulations.

It is important to note that all simulations conducted in the present section are based on the simple van der Waals model of disjoining pressure introduced in Sect. 1.7. More sophisticated approaches to modeling of disjoining pressure, discussed, e.g., in [45,72,111], can lead to modifications of some of the predictions discussed here.

3.6 Collapse of a Spherical Bubble

Dynamics of gas or vapor bubbles in liquid can be very different depending on the pressure inside the bubble. If the pressure in the gas phase is very low, the bubble is essentially a void, as envisioned, e.g., in the classical work of Rayleigh [106], and is expected to collapse very quickly. For high pressure inside the bubble, e.g., during boiling, rapid expansion is observed. The models of bubbles discussed in the previous sections of the present chapter are based on the assumption that the pressure in the gas phase inside the bubble is not dramatically different from the pressure in the liquid away from the bubble surface so that the bubble is either steady or slowly deforming. In the present section, we introduce a different class of models appropriate for describing rapidly expanding and collapsing bubbles.

In most microscale interfacial flows, the effects of inertia (described by the nonlinear terms in the Navier–Stokes equations) are negligible due to small values of the Reynolds number, as discussed in Sect. 1.5*. However, this approximation is in general not applicable to situations involving rapidly moving interfaces. Consider a spherical bubble of initial radius a in a liquid of density ρ away from solid boundaries and assume that the pressure inside the bubble is much lower than that in the liquid, so that the bubble is essentially a spherical void. The general Navier–Stokes and continuity equations in spherical coordinates (see, e.g., [1]) under the assumption of spherical symmetry reduce to

$$\frac{\partial u}{\partial t} + u\frac{\partial u}{\partial r} = -\frac{1}{\rho}\frac{\partial p}{\partial r} + \nu\left[\frac{1}{r^2}\frac{\partial}{\partial r}\left(r^2\frac{\partial u}{\partial r}\right) - \frac{2u}{r^2}\right], \qquad (3.81)$$

$$\frac{1}{r^2}\frac{\partial}{\partial r}\left(r^2 u\right) = 0, \qquad (3.82)$$

where $\nu = \mu/\rho$ is the kinematic viscosity, u is the velocity component in the radial direction. In contrast to most previous developments in the present chapter, we use *dimensional* variables here.

The interface position is described by a function $r = R(t)$ and the kinematic boundary condition is of the form

$$u\big|_{r=R(t)} = \frac{dR}{dt}. \qquad (3.83)$$

Since the pressure in the gas is negligible, the general normal stress condition at the interface, (1.82) from Sect. 1.5*, reduces to

$$\mathbf{n}\cdot\mathbf{T}\cdot\mathbf{n} = \frac{2\sigma}{R}, \qquad (3.84)$$

where $\mathbf{n}$ is the normal vector to the interface pointing into the bubble, $\mathbf{T}$ is the stress tensor in the liquid. The left-hand side of (3.84) is equal to the component T_{rr} of the stress tensor, which in turn can be expressed in terms of the local pressure and velocity as:

$$T_{rr} = -p + 2\mu \frac{\partial u}{\partial r}.$$ (3.85)

This expression allows us to rewrite (3.84) as:

$$p = -\frac{2\sigma}{R} + 2\mu \frac{\partial u}{\partial r}\bigg|_{r=R(t)}.$$ (3.86)

The pressure in the liquid far away from the bubble is known and uniform, so we write

$$p|_{r \to \infty} = p_0.$$ (3.87)

The solution of the continuity equation (3.82) is

$$r^2 u = F(t),$$ (3.88)

where $F(t)$ is an unknown function of time at this point. By substituting u from (3.88) into (3.81) we obtain

$$\frac{1}{r^2}\frac{dF}{dt} - \frac{2F^2}{r^5} = -\frac{1}{\rho}\frac{\partial p}{\partial r}.$$ (3.89)

Remarkably, the viscous terms on the right-hand side of the equation are zero regardless of the value of viscosity.

In order to obtain a simple estimate of the time it takes for the bubble to collapse, we follow Rayleigh [106] and neglect the contributions of viscous terms and surface tension in the normal stress balance, (3.86) (both of these are incorporated in the formulation of the next section). Integrating (3.89) from the current bubble radius $R = R(t)$ to infinity we obtain

$$\frac{1}{R}\frac{dF}{dt} - \frac{F^2}{2R^4} = -\frac{p_0}{\rho}.$$ (3.90)

Since (3.88) is valid everywhere in the liquid including the interface, it can be used to express the quantity $F(t)$ in terms of the interfacial velocity $V = dR/dt$ as follows,

$$F(t) = R^2(t)V(t).$$ (3.91)

Substituting this formula into (3.90) and using $\frac{dV}{dt} = V\frac{dV}{dR}$, we obtain

$$\frac{3V^2}{2} + \frac{1}{2}R\frac{dV^2}{dR} = -\frac{p_0}{\rho}.$$ (3.92)

This equation is linear in V^2 and has the solution

$$V^2 = \frac{2p_0}{3\rho}\left(\frac{a^3}{R^3} - 1\right). \tag{3.93}$$

Here, we used the initial condition of $V = 0$ at $R = a$. Now recalling that $V = dR/dt$, the radius can be expressed as a function of time in terms of an integral evaluated numerically. From this solution, we obtain the formula for the total time it takes for the bubble to collapse:

$$\tau = 0.915a\sqrt{\rho/p_0}. \tag{3.94}$$

Note that this estimate is based on a model which neglects surface tension and viscous effects, both of which can be important for bubbles of small sizes and are therefore incorporated into the more general model discussed in the next section.

3.7 The Rayleigh–Plesset Equation

The model from the previous section leads to estimates of the time of collapse of the bubble but does not address the mechanism of bubble formation. In experiments, bubbles in isothermal liquids form in the regions of low pressure. Note that this mechanism of formation, often referred to as cavitation, does not require external heating, in contrast to, e.g., bubbles seen in a boiling pot of water on a stovetop. Once a cavitation bubble is formed, it moves with the flow and thus the pressure around it changes. Instead of the simple condition of constant pressure far away from the bubble, a prescribed time-dependent pressure in the liquid $p_0(t)$ is used to model changes in pressure due to both bubble motion through the flow and possible changes in the liquid pressure specifically designed to control bubble dynamics, as is done in some experiments. To further improve the model, we have to account for the pressure inside the bubble, which although small is not identically zero when some gas or vapor is present. Assuming ideal gas law, the pressure in the bubble depends on the volume as:

$$p_B = p_{B0}(a/R)^{3k}, \tag{3.95}$$

where the constant k depends on how the temperature changes inside the bubble and is equal to unity for an isothermal process. To account for surface tension and viscosity effects at the interface, the normal stress condition given by (3.86) can be included when (3.89) is integrated from $r = R$ to infinity, resulting in the following equation for the bubble radius,

$$\frac{p_B(t) - p_0(t)}{\rho} = R\frac{d^2R}{dt^2} + \frac{3}{2}\left(\frac{dR}{dt}\right)^2 + \frac{4\nu}{R}\frac{dR}{dt} + \frac{2\sigma}{\rho R}, \tag{3.96}$$

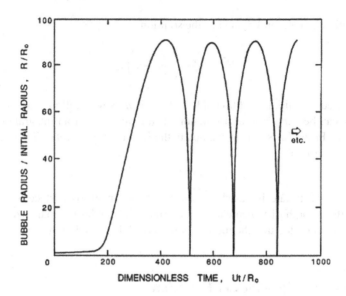

Fig. 3.15   A typical solution of the Rayleigh–Plesset equation. From [21]

often referred to as the Rayleigh–Plesset equation. Dynamics of bubble shape is thus reduced to a single differential equation which can be solved numerically. A typical numerical solution is shown in Fig. 3.15. It illustrates that the bubble can go through several cycles of rapid growth and collapse. It is important to emphasize that in contrast to most situations discussed in the present book, the inertia effects are essential for correctly capturing the fluid interface dynamics, despite the fact that the bubble size is small. Due to large characteristic interface and flow velocities, the corresponding Reynolds number Re cannot be assumed small. In situations when inertia effects are important for flows with interfaces, it is often convenient to use the nondimensional parameter called the Weber number, which characterizes the relative importance of inertia compared to surface tension. It is equal to the product of the Reynolds number and the capillary number Ca.

The rapid dynamics of bubbles in regions of low pressure provides a qualitative explanation for wear to turbines and ship propellers known as cavitation damage. However, in order to develop a quantitatively accurate description of cavitation near solid surfaces it is important to take into account that the presence of solid walls can make the bubble collapse highly asymmetric, as discussed in Brennen [21]. Models assuming spherical symmetry do not describe this situation.

A remarkable set of phenomena related to bubble collapse is sonoluminescence, when light is emitted by collapsing microbubbles. A review on the models of this phenomenon is provided by, e.g., Brenner et al. [22].

3.8 Notes on Literature

Models of confined bubbles and their applications to microscale flows are reviewed by Guenther and Jensen [61] and Ajaev and Homsy [5]. The mathematical theory of equilibrium capillary surfaces is the subject of the well-known monograph of Finn [53]. Slow motion and interaction of bubbles in large volumes of liquid away from confining walls is beyond the scope of this book, but was discussed, e.g., in several books [87, 88, 125]. The drainage problem, as well as many other important topics in interfacial fluid mechanics, is discussed by Slattery et al. [115]. A wide range of problems related to dynamics of rapidly deforming bubbles is discussed in the book of Brennen [21].

Chapter 4
Flows in the Presence of Electric Charges and Fields

4.1 Electrical Double Layer

Electric charges at solid–liquid and liquid–gas interfaces can affect flow in the liquid and evolution of liquid–gas interfaces. Consider, for example, a gas bubble approaching a solid surface as discussed in Chap. 3. The model formulated there does not account for any electric charges and describes interface deformation under the action of surface tension and, for sufficiently small film thickness, London–van der Waals forces. If there are electric charges at the interfaces, electrostatic interactions can have a significant effect on the dynamics of the bubble. For example, if the charges are of the same sign, the bubble surface and the solid wall will repel each other and the liquid film between them may never approach the range of thicknesses at which the London–van der Waals forces become important. Thin liquid films that would otherwise rupture can be stabilized by the electrostatic repulsion.

Electric charges are also often present in liquids away from interfaces. In water, ions are always present due to contamination and, even for deionized pure water, due to self-ionization of H_2O molecules. Concentration of ions can be varied over several orders of magnitude in ionic solutions, such as a solution of table salt in water which contains the ions of sodium (Na^+) and chloride (Cl^-). The superscripts $+$ and $-$ are used for ions of electric charges of e and $-e$, respectively; e is the elementary charge. Ionic solutions often contain several types of ions, numbered below by an integer i, possibly of different valencies z_i. In the present context, valence is simply an integer multiplying e in the formula for the electrical charge of an ion. For example, ions of calcium Ca^{2+} have an electric charge of $2e$ and their valence is equal to 2.

A solid–liquid interface can acquire electric charge by adsorbing ions present in the liquid or by ionization of the molecules of the solid. An example of the latter mechanism is ionization of common glass in contact with water, described by the chemical equation

$$SiOH \rightarrow SiO^- + H^+. \tag{4.1}$$

V.S. Ajaev, *Interfacial Fluid Mechanics: A Mathematical Modeling Approach*,
DOI 10.1007/978-1-4614-1341-7_4, © Springer Science+Business Media, LLC 2012

Fig. 4.1 Sketch illustrating
the mechanism of formation
of an electrical double layer
near a positively charged
solid wall

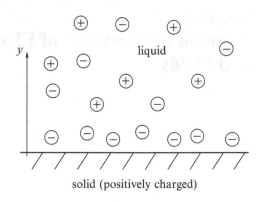

solid (positively charged)

It results in appearance of negatively charged surface groups SiO^-. The equilibrium value of the surface charge can be estimated by considering all relevant processes of ionization and ion adsorption and turns out to be negative for glass surface in water under most experimental conditions. In general, solid surfaces in contact with liquids can carry either positive or negative charge.

Consider a flat solid surface carrying a positive electric charge of uniform density. It is clear that negatively charged ions in the liquid (called anions) will be attracted to the solid and positively charged ones (cations) repelled from it, but thermal motion will counteract these tendencies. As a result of the balance between these two physical effects, a certain equilibrium distribution of electric charges will be established near the solid, with excess negative charge closer to the solid–liquid interface, as schematically shown in Fig. 4.1. The liquid is assumed to be electrically neutral away from boundaries, so the density of electric charges will be nonzero only in a layer near the solid–liquid boundary, called the *electrical double layer*. The electric field in the liquid near the solid can be characterized by its potential ψ which for uniform surface charge density is a function of only one Cartesian coordinate, normal to the wall, denoted by y. Assuming thermodynamic equilibrium, the electrochemical potential (defined in Appendix A) for each species is constant throughout the domain of the liquid phase. In the calculation below, we assume that μ_i in the definition of the electrochemical potentials are approximated by the chemical potentials of species in an ideal solution [79]. Then, the condition of constant electrochemical potential in terms of ion concentrations n_i (number of ions per unit volume) and the electric potential can be written as:

$$k_B T \ln n_i + z_i e \psi = \text{const.} \qquad (4.2)$$

or, equivalently

$$n_i = n_i^{(0)} \exp\left(-\frac{z_i e \psi}{k_B T}\right). \qquad (4.3)$$

Here, k_B is the Boltzmann constant, T is temperature, and $n_i^{(0)}$'s denote ion concentrations under the condition of $\psi \to 0$, which is usually far away from the solid surface.

The electric field in a liquid with an electric charge density ρ_E satisfies the classical Poisson's equation,

$$\nabla^2 \psi = -\frac{\rho_E}{\varepsilon}, \tag{4.4}$$

where ε is the dielectric permittivity. For water, $\varepsilon \approx 80\varepsilon_0$, where ε_0 is the permittivity of the vacuum. We use the standard SI units here, although many studies of electrokinetic phenomena are based on equations in cgs units; a useful conversion table between the key equations in the two systems of units can be found in Chap. 5 of Adamson and Gast [2].

The electric charge density ρ_E on the right-hand side of (4.4) can be expressed in terms of ion concentrations. In particular, when ion concentration follows the Boltzmann distribution (4.3), the so-called *Poisson–Boltzmann* equation is obtained:

$$\nabla^2 \psi = -\sum_i \frac{z_i e n_i^{(0)}}{\varepsilon} \exp\left(-\frac{z_i e \psi}{k_B T}\right). \tag{4.5}$$

In the present case, ψ is a function of only one spatial coordinate, so we can replace $\nabla^2 \psi$ with $d^2\psi/dy^2$. The boundary conditions for the potential are usually written in the form

$$\psi = \psi_0 \quad \text{at} \quad y = 0, \tag{4.6}$$

$$\psi \to 0 \quad \text{as} \quad y \to \infty, \tag{4.7}$$

where the solid surface potential ψ_0 is assumed to be fixed. Alternatively, one can specify the electric charge density at the solid.

The right-hand side of (4.5) can be simplified for an important special case of a liquid which is electrically neutral away from the solid surface and contains only two types of ions, so that $i = 1, 2$ with $n_1^{(0)} = n_2^{(0)} = n_0$, $z_1 = -z_2 = z$ (the so-called *symmetric electrolyte*). The one-dimensional version of the Poisson–Boltzmann equation for this case is

$$\frac{d^2\psi}{dy^2} = \frac{2ezn_0}{\varepsilon} \sinh\left(\frac{ez\psi}{k_B T}\right). \tag{4.8}$$

By introducing a new variable $\eta = ez\psi/k_B T$, we can rewrite this equation as:

$$\frac{d^2\eta}{dy^2} = \lambda_D^{-2} \sinh \eta, \tag{4.9}$$

where λ_D is the Debye length,

$$\lambda_D = \sqrt{\frac{\varepsilon k_B T}{2 n_0 e^2 z^2}}. \tag{4.10}$$

This quantity is the characteristic length of change of the electric field near the solid surface. It arises in several other contexts, most notably in plasma physics, in situations when rearrangement of mobile charge carriers results in modification (or *screening*) of an electric field. The Debye length is between several angstroms and $\sim 100\,\mathrm{nm}$ for typical ionic solutions used in applications and of the order of $1\,\mu\mathrm{m}$ for pure water. Introducing a nondimensional coordinate $\tilde{y} = y/\lambda_D$, the equation for η can be written in the form

$$\frac{d^2 \eta}{d \tilde{y}^2} = \sinh \eta. \tag{4.11}$$

This ordinary differential equation can be solved analytically by observing that the second derivative of η can be expressed in terms of $\zeta(\eta) = d\eta/d\tilde{y}$ as follows,

$$\frac{d^2 \eta}{d \tilde{y}^2} = \zeta \frac{d\zeta}{d\eta} = \frac{1}{2} \frac{d\zeta^2}{d\eta}. \tag{4.12}$$

Equation (4.11) then becomes a first-order equation for $\zeta(\eta)$. Integrating this equation in η with the condition $\zeta(0) = 0$, which simply reflects the fact that both η and $d\eta/d\tilde{y}$ decay to zero as $\tilde{y} \to \infty$, we obtain

$$\frac{d\eta}{d\tilde{y}} = -\sqrt{2(\cosh \eta - 1)}. \tag{4.13}$$

The decay condition for η away from the wall requires the sign in front of the square root to be negative (assuming $\psi_0 > 0$). Using properties of hyperbolic functions, (4.13) can be written as:

$$\frac{d\eta}{d\tilde{y}} = -2 \sinh \frac{\eta}{2} \tag{4.14}$$

and then integrated (using $\sinh \frac{\eta}{2} = 2 \sinh \frac{\eta}{4} \cosh \frac{\eta}{4}$) to give

$$\eta = 4 \tanh^{-1} \left(e^{-\tilde{y}} \tanh \frac{\eta_0}{4} \right), \tag{4.15}$$

where $\eta_0 = e z \psi_0 / k_B T$ is the scaled electric potential of the solid surface. This result is often written in terms of exponentials

$$\eta = 2 \ln \frac{1 + \alpha_0 e^{-\tilde{y}}}{1 - \alpha_0 e^{-\tilde{y}}}, \qquad \alpha_0 = \frac{e^{\eta_0/2} - 1}{e^{\eta_0/2} + 1}, \tag{4.16}$$

and referred to as the Gouy–Chapman solution of the Poisson–Boltzmann equation.

Equation (4.5) is in general impossible to solve analytically when several types of ions are present in the liquid. However, if $z_i e \psi / k_B T \ll 1$, it can be linearized, resulting in the so-called Debye–Hückel equation:

$$\nabla^2 \psi = \lambda_D^{-2} \psi, \tag{4.17}$$

where the definition of the Debye length is now generalized to include contributions of several types of ions,

$$\lambda_D = \left(\frac{e^2}{\varepsilon k_B T} \sum_i n_i^{(0)} z_i^2 \right)^{-1/2}. \tag{4.18}$$

In deriving (4.17), we assumed that the liquid is electrically neutral far away from the solid surface and therefore $\sum_i z_i n_i^{(0)} = 0$. Solving (4.17) with the boundary conditions (4.6)–(4.7) results in

$$\psi = \psi_0 \exp \left(-\frac{y}{\lambda_D} \right). \tag{4.19}$$

Thus, the Debye–Hückel equation predicts that the potential in the liquid in the electrical double layer decays exponentially with the characteristic decay length of λ_D.

To illustrate limitations of the linearized model expressed by (4.17), let us compare its predictions for the case of a symmetric electrolyte $\left(n_1^{(0)} = n_2^{(0)}, z_1 = -z_2 = 1 \right)$ with the known analytical solution for this case, given by (4.15). Using the same scaled variables as in (4.15), the linearized solution (4.19) is written as:

$$\eta = \eta_0 \exp(-\tilde{y}). \tag{4.20}$$

While the linearization of the Poisson–Boltzmann equation is strictly speaking only valid for $\eta_0 \ll 1$, the results obtained from (4.15) and (4.20) are in reasonable agreement up to $\eta_0 \sim 1$. However, as η_0 is increased further, discrepancies between the two solutions are clearly seen, as illustrated in Fig. 4.2 for a value of the surface potential corresponding to η_0 near 4. The Debye–Hückel equation predicts higher values of the scaled potential than the analytical Gouy–Chapman solution of the full Poisson–Boltzmann equation.

Once the electric potential in the double layer is determined as a function of the distance to the wall, the distribution of ions can be found using (4.3). For a negative z_i and a typical wall potential of 0.3 V, the value of n_i turns out to be several orders of magnitude higher than the bulk concentration $n_i^{(0)}$, which in dimensional terms leads to physically unrealistic values. To address this shortcoming, we note that the effect of finite size of ions (which so far have been assumed to be point charges) can be important in a thin sublayer of the electrical double layer very close to the wall, often called the Stern layer [2, 102]. Ion adsorption takes place in this region and

Fig. 4.2 Electric potential in the liquid as a function of the distance to the charged wall in scaled variables, as predicted by the Gouy–Chapman solution (*solid line*) and the Debye–Hückel equation with $n_1^{(0)} = n_2^{(0)}, z_1 = -z_2 = 1$ (*dot-dashed line*). The dimensional solid surface potential is 0.1 V, temperature is 25°C

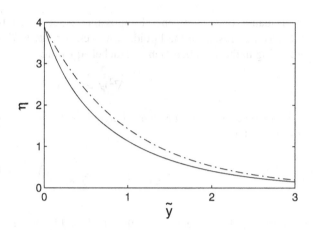

the Boltzmann distribution of ion concentrations, (4.3), is no longer valid. Liquid in the Stern layer and close to it can be considered immobile; the plane separating this part from the mobile portion of the electrical double layer is called the shear surface. The electric potential at the shear surface is denoted by ζ and is often called the ζ-potential. Since the Stern layer thickness is usually much smaller than λ_D, its effect on the solution in the electrical double layer can be accounted for by simply replacing the boundary condition (4.6) with

$$\psi = \zeta \quad \text{at} \quad y = 0. \tag{4.21}$$

The solution of the Debye–Hückel equation with this boundary condition is $\psi = \zeta e^{-y/\lambda_D}$, so the ζ-potential can also be interpreted as the electric potential drop across the screening cloud of mobile ions distributed according to (4.3).

4.2 Electroosmotic Flow

The Boltzmann distribution of ion concentrations near a charged wall, (4.3), and solutions for the electric potential discussed in the previous section, imply that the electric charge density,

$$\bar{\rho}_E = \sum_i z_i e n_i^{(0)} \exp\left(-\frac{z_i e \psi}{k_B T}\right), \tag{4.22}$$

is nonzero in the region of the characteristic width of λ_D. This opens up a possibility to induce and control liquid motion by applying external electric fields. Electrically induced motion of liquid with respect to a charged surface is called *electroosmotic flow*. Assuming the conditions of applicability of the Stokes flow model, introduced

Fig. 4.3 Sketch illustrating the velocity profile near a uniformly charged solid surface for a steady electroosmotic flow generated by an external electric field E_{ext} parallel to the surface

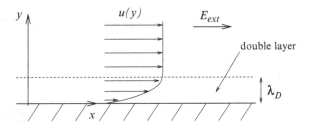

in Sect. 1.5*, are satisfied, the dimensional pressure p and velocity field $\mathbf{u}$ are described by:

$$\nabla p = \mu \nabla^2 \mathbf{u} + \rho_E \mathbf{E}, \tag{4.23}$$

$$\nabla \cdot \mathbf{u} = 0, \tag{4.24}$$

where in the expression for the body force we neglect contributions from gravity, but include a term due to the electric field, $\mathbf{E}$. The latter is in general a superposition of an applied external field, of magnitude E_{ext}, and the electric field due to ions in the electrical double layer. Note that once an external field is applied, the charge density, ρ_E, in the equation for fluid flow is in general not the same as $\bar{\rho}_E$ from (4.22), due to rearrangement of ions under the action of the external field. However, for an important special case of an external field parallel to a uniformly charged solid surface, as sketched in Fig. 4.3, the distribution of ions is not affected by the external field and therefore we can assume $\rho_E = \bar{\rho}_E$. Furthermore, assuming no external pressure gradient and no variations of the velocity and pressure fields in the direction of the external electric field (chosen as the direction of the x-axis), the Stokes flow equations in coordinate form are reduced to

$$0 = \mu u_{yy} + \bar{\rho}_E E_{ext}, \tag{4.25}$$

$$p_y = \mu v_{yy} - \bar{\rho}_E \psi_y, \tag{4.26}$$

$$v_y = 0, \tag{4.27}$$

where u and v are the velocity components in the x- and y-directions, respectively. The y-axis is normal to the wall and the model is two-dimensional, i.e., no variation is assumed in the direction normal to the plane of the sketch in Fig. 4.3. As discussed in the end of the previous section, the Stern layer, which is essentially immobile, is typically much thinner that the electrical double layer, so instead of incorporating a finite layer of immobile liquid into our formulation we simply apply the boundary conditions of zero velocity at the solid–liquid boundary itself,

$$u = v = 0 \quad \text{at} \quad y = 0. \tag{4.28}$$

The no-penetration condition ($v = 0$) together with (4.27) immediately implies that the vertical velocity component is zero everywhere and therefore the pressure distribution in the electrical double layer satisfies

$$p_y = -\sum_i z_i e n_i^{(0)} \exp\left(-\frac{z_i e \psi}{k_B T}\right) \psi_y. \tag{4.29}$$

Here we used the expression for the charge density from (4.22). Integrating (4.29) with the conditions of $p \to p_0$ and zero electric potential far away from the wall, we obtain

$$p = p_0 + \sum_i k_B T n_i^{(0)} \left[\exp\left(-\frac{z_i e \psi}{k_B T}\right) - 1\right]. \tag{4.30}$$

In order to find the horizontal velocity profile from (4.25), it is convenient to express the charge density in terms of ψ using the Poisson's equation, $\varepsilon \nabla^2 \psi = -\bar{\rho}_E$, so that (4.25) becomes

$$0 = \mu u_{yy} - \varepsilon \psi_{yy} E_{ext}, \tag{4.31}$$

and can then be easily integrated in y, resulting in

$$u = \frac{\varepsilon \psi E_{ext}}{\mu} + c_1 y + c_2. \tag{4.32}$$

Since the velocity is bounded away from the wall (at $y \to \infty$), the constant c_1 has to be zero. Once the value of c_2 is determined by combining the conditions (4.21) and (4.28), the velocity profile can be written as:

$$u = \frac{\varepsilon E_{ext}}{\mu}(\psi - \zeta). \tag{4.33}$$

We note that the derivation of this equation did not rely on the assumption of $z_i e \psi / (k_B T)$ being small, so the applicability of the linear relation between the velocity and the potential in (4.33) is not limited to small electric potentials. Based on (4.33), the velocity of liquid motion changes from zero to

$$u_s = -\frac{\varepsilon E_{ext} \zeta}{\mu} \tag{4.34}$$

in the layer of the characteristic thickness equal to the Debye length, which is the length scale of variation of ψ, as was determined in the previous section. Since λ_D is much smaller than other relevant dimensions in typical microscale applications, the effect of electrical double layer is often represented by replacing the classical no-slip condition at the solid wall with the condition of $u = u_s$, called the Helmholtz–Smoluchowski slip condition. This neglects the velocity gradient within the double layer but produces a velocity profile accurate in regions away from the wall. The directions of the flow velocity and the electric field are the same for negative ζ (corresponding to negatively charged wall), as sketched in Fig. 4.3, and are opposite for positive ζ.

In the discussion of liquid transport on microscale in Sect. 1.3, we noted that common approaches to liquid transport become less efficient as the system size decreases. In particular, for pressure-driven flow in a long microchannel, the two-dimensional model developed in Sect. 1.3 predicted the average flow velocity to decrease as the square of the height of the channel. Let us now suppose that channel walls are electrically charged and instead of using pressure differences to generate the flow, we apply an electric field in the direction parallel to the channel walls. Since the Debye length is typically much smaller than the channel height, the velocity will be nearly uniform and equal to u_s throughout the channel cross-section, meaning that the average value is

$$\bar{u} = -\frac{\varepsilon E_{ext} \zeta}{\mu}. \tag{4.35}$$

Clearly, this average velocity does not decrease as the system size (in the present case the channel height) is reduced, which is in contrast to the average velocities for all previously considered transport mechanisms, such as the value determined by (1.31). Thus, electroosmotic flow is an efficient method for transporting liquids on microscale. This motivated significant amount of research on the topic, reviewed, e.g., in Stone et al. [123].

The improvement in the transport rate at small scales is not the only advantage of electroosmotic flow compared to pressure-driven flow. In many practical applications of microfluidics, a specific amount of dissolved chemical, e.g., macromolecules, has to be transported by the liquid through a microchannel. For the parabolic velocity profile discussed in Sect. 1.3, different layers of liquid move at different velocities, so if the chemical sample to be transported is initially highly localized, it will undergo significant broadening as is travels through the channel, an effect known as hydrodynamic dispersion. For electroosmotic flow, the flow velocity is uniform except in a thin layer near the wall, so it is much easier to avoid hydrodynamic dispersion while transporting localized regions of high concentration of dissolved chemicals.

The assumption of uniform electric charge at the solid surface is often violated in microscale devices due to adsorption of chemical species or surface defects, resulting in more complicated flow patterns than discussed in the present section [57]. Furthermore, alternating oppositely charged patches of solid surface are used in some applications, e.g., to improve mixing at small scales [120, 123]. Many features of electroosmotic flow near surfaces with nonuniform charge distribution can be captured by models using spatially dependent rather than constant ζ-potential as the boundary condition at the solid surface.

4.3   Electrostatic Component of Disjoining Pressure

4.3.1   Equilibrium Conditions for Liquid Film on Charged Surface

When a gas bubble is pressed against a solid wall in the experimental configuration introduced in Chap. 3, the flattened portion of the bubble surface near the wall can be in equilibrium with the curved meniscus away from it. For nonpolar liquids in the absence of electric charges, this phenomenon is explained by the action of the London–van der Waals dispersion forces, discussed in detail in Sect. 1.7. However, it is important to emphasize that in the presence of ions and electric charges at interfaces, the electrostatic effects can play a more important role in determining the interface shapes than the dispersion forces. In order to describe these effects, consider a thin liquid film region formed between the gas bubble and the solid wall, as sketched in Fig. 4.4. While the interface is axisymmetric in experiments, we consider a two-dimensional version of the problem here since the derivation for this simpler configuration is essentially the same as for the axisymmetric case. The dimensional Cartesian coordinates x and y are along the solid surface and normal to it, respectively. For a liquid at rest and in the absence of external electric fields, the momentum equations for liquid flow reduce to

$$\nabla p + \sum_i z_i e n_i^{(0)} \exp\left(-\frac{z_i e \psi}{k_B T}\right) \nabla \psi = 0. \tag{4.36}$$

Here, we used the general formula for charge density expressed in terms of the local electric potential, (4.22). Equation (4.36) has a simple physical meaning: it expresses the balance between liquid pressure and electrostatic forces required for liquid to be at rest. By introducing

$$P = p - \sum_i n_i^{(0)} k_B T \exp\left(-\frac{z_i e \psi}{k_B T}\right), \tag{4.37}$$

one can write (4.36) in the form $\nabla P = 0$, which immediately implies that the function P is constant throughout the liquid domain. Thus, the value of P at an arbitrary point A in the film, P_A, is equal to the value P_B in the bulk of the liquid.

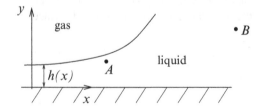

Fig. 4.4 Sketch of a thin liquid film at rest on a charged solid surface

The latter is easily expressed in terms of the bulk pressure p_0 since the electric field there can be neglected as long as the point B is sufficiently far away from the wall, so the pressure p_A at an arbitrary point A in the liquid film can be expressed in terms of the potential there, ψ_A, as:

$$p_A = p_0 + \sum_i n_i^{(0)} k_B T \left[\exp \left(-\frac{z_i e \psi_A}{k_B T} \right) - 1 \right]. \tag{4.38}$$

The pressure difference between the film and the bulk liquid due to electrostatic effects, $p_A - p_0$, with p_A evaluated near the liquid–gas interface, is usually referred to as the *electrostatic component of disjoining pressure* and denoted by Π_e. In principle, both this component and the van der Waals contribution to disjoining pressure, given by (1.129), can affect stresses in thin liquid films. A simple superposition of these two different physical effects is often used and referred to as the DLVO theory of disjoining pressure, after Derjaguin, Landau, Verwey, and Overbeek, although these authors in fact considered a related problem of interaction of two solid surfaces through a liquid medium.

In order to calculate the electric potential in the liquid film (needed for finding the value of p_A), we assume the characteristic length of change of the fluid interface shape in the x-direction to be much larger than that in the y-direction. Then, the Poisson–Boltzmann equation, introduced in Sect. 4.1, can be written as:

$$\psi_{yy} = -\sum_i \frac{z_i e n_i^{(0)}}{\varepsilon} \exp \left(-\frac{z_i e \psi}{k_B T} \right). \tag{4.39}$$

Assuming the solid wall to be at a uniform potential ψ_0 with no external electric field and neglecting the effects of the Stern layer near the wall, the boundary conditions for this equation are formulated as:

$$\psi = \psi_0 \quad \text{at} \quad y = 0, \tag{4.40}$$

$$\psi_y = 0 \quad \text{at} \quad y = h. \tag{4.41}$$

The second condition follows from continuity of the electric field at the liquid–air interface

$$\varepsilon \psi_y^- = \varepsilon_{\text{air}} \psi_y^+ \quad \text{at} \quad y = h, \tag{4.42}$$

in the limit of $\varepsilon \gg \varepsilon_{\text{air}} \approx \varepsilon_0$, where the superscripts denote one-sided derivatives, ε_{air} is the dielectric permittivity of air. For water, the ratio $\varepsilon / \varepsilon_0$ is near 80.

4.3.2   Disjoining Pressure for Symmetric Electrolytes

Equation (4.39) is nonlinear and in general has to be solved numerically. However, some analytical results can be obtained for the important special case of a symmetric electrolyte, first discussed in Sect. 4.1. Using the notation introduced

Fig. 4.5 Sketch illustrating
the basic idea of the method
of images in electrostatics

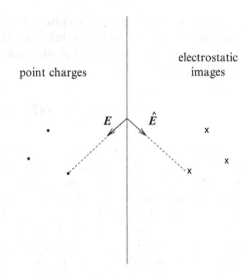

there ($\tilde{y} = y/\lambda_D$, $\eta = ez\psi/k_B T$, $z_1 = -z_2 = z$), the boundary value problem for the
scaled electric potential in the liquid film can be reformulated as:

$$\frac{d^2\eta}{d\tilde{y}^2} = \sinh\eta, \tag{4.43}$$

$$\eta(0) = \eta_0, \qquad \frac{d\eta}{d\tilde{y}}(\tilde{h}) = 0, \tag{4.44}$$

where $\tilde{h} = h/\lambda_D$. The condition of zero normal derivative of the potential at the
flat boundary suggests that the ideas of the classical method of images from
electrostatics can be used to determine the electric potential in the liquid film. We
first briefly review the basic idea of the method and then discuss how it can be
applied to the calculation of the electrostatic component of disjoining pressure.

Consider several stationary point charges in a continuous medium, as sketched
in Fig. 4.5, and assume that the electric field (equal to the negative gradient of the
electric potential) has zero normal component at the flat boundary of the medium,
shown by the solid line in the sketch. Then the electric potential in the medium is the
same as the combined potential of the charges and their electrostatic mirror images
(of the same electric charge) with respect to the solid boundary. To verify this, we
note that the combined potential satisfies the Poisson's equation ($\varepsilon\nabla^2\psi = -\rho_E$)
since no charges are added in the physical domain and has zero normal derivative
at the boundary. To prove the latter point, we observe that the normal component of
the electric field from each actual point change is balanced at the boundary by that
of its electrostatic image. This is illustrated in Fig. 4.5 by showing the vector of the
electric field from one of the charges, **E**, and its image, **Ê**; their vector sum is clearly
directed along the boundary. Thus, the combined potential satisfies the boundary
condition of zero normal derivative.

The method of images does not allow one to find the electric potential in the film of arbitrary nondimensional thickness $\tilde{h}$ since the positions of the electric charges are not known in advance and have to be determined as part of the solution, which is reflected in the fact that (4.43) is nonlinear. However, the idea of the method can be used to obtain an approximate solution valid for large values of $\tilde{h}$. Consider the solution of (4.43) obtained in Sect. 4.1, now denoted by H,

$$H = 4\tanh^{-1}\left(e^{-\tilde{y}}\tanh\frac{\eta_0}{4}\right),\tag{4.45}$$

which clearly satisfies the boundary condition at the solid–liquid but *not* at the liquid–gas interface. Suppose $\hat{H}$ is defined by:

$$\hat{H} = 4\tanh^{-1}\left(e^{\tilde{y}-2\tilde{h}}\tanh\frac{\eta_0}{4}\right).\tag{4.46}$$

The sum $\eta = H + \hat{H}$ satisfies the condition of zero normal derivative at $\tilde{y} = \tilde{h}$. If, in addition, the value of the scaled electric potential at the liquid–gas interface, $\bar{\eta} \equiv \eta(\tilde{h})$, is small then $\hat{H} \ll 1$ everywhere in the domain $[0,\tilde{h}]$ and the right-hand side of (4.43) can be simplified according to

$$\sinh(H+\hat{H}) = \sinh H \cosh\hat{H} + \cosh H \sinh\hat{H} \approx \sinh H.\tag{4.47}$$

Thus, in the limit of $\bar{\eta} \to 0$, the function η satisfies (4.43) (note also that $\hat{H}'' \ll H''$ for $0 \le \tilde{y} \le \tilde{h}$). The value of η at the solid–liquid boundary is not equal to η_0, but the difference is approaching zero in the limit of small $\bar{\eta}$. Thus, we obtained an approximate solution of the problem (4.43)–(4.44) and can now use this solution to determine the value of the scaled potential at the liquid–gas interface:

$$\bar{\eta} = 8\tanh\left(\frac{\eta_0}{4}\right)e^{-\tilde{h}}.\tag{4.48}$$

We note that since the electric potential decays away from the wall, the assumption of small $\bar{\eta}$ used in the derivation of (4.48) has a wider range of applicability than the assumption of small η_0 (corresponding to the classical Debye–Hückel approximation).

Once the electric potential near the liquid–gas interface is found, it can be used to determine the pressure p there using (4.38). For a liquid with two types of oppositely charges ions, this equation (applied at a point near the liquid–gas interface) can be written in nondimensional form as:

$$\tilde{p} = 2(\cosh\bar{\eta} - 1),\tag{4.49}$$

where $\tilde{p} = (p - p_0)/n_0 k_B T$. Using (4.48), the scaled pressure difference at the leading order in $\bar{\eta}$ is expressed as:

$$\tilde{p} = 64 \tanh^2 \left(\frac{\eta_0}{4} \right) e^{-2\tilde{h}}. \tag{4.50}$$

It is also instructive to rewrite the result in dimensional form

$$p - p_0 = 64 n_0 k_B T \tanh^2 \left(\frac{ez\psi_0}{4k_B T} \right) e^{-2h/\lambda_D}. \tag{4.51}$$

Consider an aqueous NaCl solution at 25°C with molar concentration of 1 mM (here "M" denotes a commonly used unit of molar concentration, mol/L). For the dimensional value of $\psi_0 = -122\,\mathrm{mV}$, corresponding to quartz–aqueous solution interface in the experiments of Hewitt et al. [65], and the film thickness of 30 nm, the pressure difference calculated from (4.51) is 211 Pa. This is about the same as the capillary pressure jump at the surface of a spherical air bubble of radius 0.7 mm formed in water.

Since the approximate formula for $\bar{\eta}$, (4.48), is only valid in the limit of $\bar{\eta} \to 0$, it may seem that its applicability is limited to unrealistically large values of the scaled thickness, for which the effects of disjoining pressure are not significant. To verify that this is not the case, let us recall the general nonlinear boundary value problem for η defined by (4.43)–(4.44). Following the procedure outlined in Sect. 4.1, we integrate (4.43) once in $\tilde{y}$ to obtain

$$\frac{d\eta}{d\tilde{y}} = -\sqrt{2(\cosh \eta - \cosh \bar{\eta})}, \tag{4.52}$$

where the boundary condition of zero electric field at $\tilde{y} = \tilde{h}$ has been used to determine the constant of integration and the choice of sign in front of the square root is due to the assumption that η_0 is positive. Using the condition $\eta(0) = \eta_0$, (4.52) leads to

$$\int_{\bar{\eta}}^{\eta_0} \frac{d\eta}{\sqrt{2(\cosh \eta - \cosh \bar{\eta})}} = \tilde{h}. \tag{4.53}$$

The integral on the left-hand side of this equation can be expressed in terms of elliptic integrals by observing that

$$2(\cosh \eta - \cosh \bar{\eta}) = e^{\eta} - e^{\bar{\eta}} + e^{-\eta} - e^{-\bar{\eta}} = \left(1 - e^{\bar{\eta}-\eta} \right) \left(e^{\eta} - e^{-\bar{\eta}} \right)$$
$$= e^{\bar{\eta}} \left(1 - e^{\bar{\eta}-\eta} \right) \left(e^{\eta-\bar{\eta}} - e^{-2\bar{\eta}} \right) \tag{4.54}$$

and introducing a variable $\phi = \sin^{-1} \exp[(\bar{\eta} - \eta)/2]$, defined for all η in the domain of integration since $\eta \geq \bar{\eta}$ there. Note that $d\eta = -2 \cot \phi \, d\phi$ and the integration bounds in terms of the new variable are from $\pi/2$ to $\phi_0 \equiv \sin^{-1} \exp[(\bar{\eta} - \eta_0)/2]$. Equation (4.53) can then be written as:

$$2e^{-\frac{\bar{\eta}}{2}} \int_{\phi_0}^{\pi/2} \frac{d\phi}{\sqrt{1 - e^{-2\bar{\eta}} \sin^2 \phi}} = \tilde{h} \tag{4.55}$$

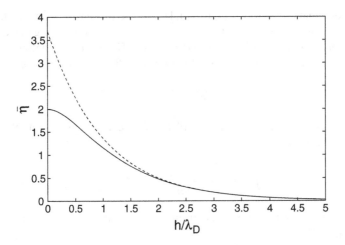

Fig. 4.6 The scaled electric potential at the liquid–gas interface as a function of the scaled film thickness found from the nonlinear Poisson–Boltzmann equation (*solid line*) and from the small $\tilde{\eta}$ approximation (*dashed line*)

or, equivalently, as:

$$2e^{-\frac{\tilde{\eta}}{2}}\left[K\left(e^{-\tilde{\eta}}\right) - F\left(\phi_0, e^{-\tilde{\eta}}\right)\right] = \tilde{h}. \tag{4.56}$$

Here, we used the standard notation for the incomplete elliptic integral of the first kind,

$$F(\phi, k) = \int_0^\phi \frac{d\theta}{\sqrt{1 - k^2 \sin^2 \theta}}, \tag{4.57}$$

and the complete elliptic integral $K(k) = F(\pi/2, k)$. The scaled electric potential at the interface, $\tilde{\eta}$, can be obtained from the implicit relation (4.56) as a function of $\tilde{h}$ using, e.g., MATLAB or *Mathematica*. The result is shown by the solid line in Fig. 4.6. The dashed line in the figure corresponds to the approximate result given by (4.48). The agreement between the two is reasonable even for $h \sim \lambda_D$ and excellent for the film thickness above $2\lambda_D$. For small $\tilde{h}$, the approximate model predicts significantly higher values of the electric potential than the solution of the full nonlinear boundary value problem given by (4.43)–(4.44).

4.3.3   Debye–Hückel Approximation

For the more general case of a liquid with several different types of ions, the electric potential in a uniform film can in principle be found by solving the one-dimensional nonlinear Poisson–Boltzmann equation, (4.39), numerically on the interval $[0, h]$

with the boundary conditions (4.40)–(4.41). In the present section, we limit our discussion to the case when $|z_i e \psi_0 / k_B T| \ll 1$ for all z_i, corresponding to the Debye–Hückel approximation first introduced in Sect. 4.1. The electric potential in this case is found from the linearized version of the Poisson–Boltzmann equation,

$$\psi_{yy} = \lambda_D^{-2} \psi, \tag{4.58}$$

where the Debye length is defined by (4.18). It is convenient to write the general solution of (4.58) as:

$$\psi = c_1 \cosh \frac{y - h}{\lambda_D} + c_2 \sinh \frac{y - h}{\lambda_D}. \tag{4.59}$$

The constants c_1 and c_2 are found from the boundary conditions (4.40)–(4.41), resulting in

$$\psi = \frac{\psi_0}{\cosh(h/\lambda_D)} \cosh \frac{y - h}{\lambda_D}. \tag{4.60}$$

The electrostatic component of disjoining pressure can be expressed in terms of the electric potential in the film by expanding (4.38) in powers of the small parameters $z_i e \psi_0 / k_B T$ and considering only the leading-order nonzero terms in the expansion. Since $\sum_i n_i^{(0)} z_i = 0$ (liquid is electrically neutral in the bulk), the terms linear in ψ in the expansion add up to zero and the disjoining pressure is approximated by:

$$p - p_0 = \frac{1}{2} \varepsilon \lambda_D^{-2} \psi^2. \tag{4.61}$$

Here we used the definition of the Debye length from (4.18). Evaluating p at the liquid–gas interface and using (4.60), we obtain

$$p - p_0 = \frac{\varepsilon \lambda_D^{-2} \psi_0^2}{2 \cosh^2(h/\lambda_D)}. \tag{4.62}$$

It is important to note that (4.60) is applicable for all values of the film thickness as long as the electric potential at the wall is small enough for the Debye–Hückel approximation to be valid. We illustrate this for the special case of a symmetric electrolyte $\left(n_1^{(0)} = n_2^{(0)} = n_0, \ z_1 = -z_2 = 1 \right)$ for which the exact value of the scaled potential at the liquid–gas interface is given implicitly by (4.56). The numerical result obtained from this formula for $\eta_0 = 0.5$ is shown by the solid line in Fig. 4.7. The value of the electric potential at the liquid–gas interfaces, $\psi(h)$ from (4.60) expressed in the nondimensional terms as:

$$\bar{\eta} = \frac{\eta_0}{\cosh \bar{h}}, \tag{4.63}$$

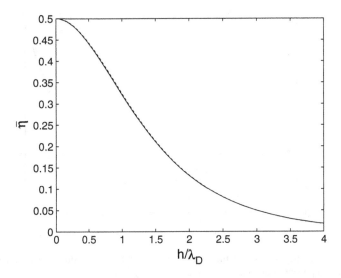

Fig. 4.7 The scaled electric potential at the liquid–gas interface as a function of the scaled film thickness found from the numerical solution of (4.56), *solid line*, and from the Debye–Hückel approximation, *dashed line*

is shown by the dashed line in Fig. 4.7. Even for the relatively large value of the scaled wall potential, $\eta_0 = 0.5$, the agreement between the two is excellent. The MATLAB code used for generating Fig. 4.7 can be found in Sect. B.4.

Finally, for the film thickness much larger than the Debye length, (4.62) simplifies to

$$p - p_0 = 2\varepsilon \lambda_D^{-2} \psi_0^2 e^{-2h/\lambda_D}. \tag{4.64}$$

To provide a brief summary of the analytical formulas used for calculating the electrostatic component of disjoining pressure, we note that for the general case of liquid with several types of ions, analytical formulas are available only when $|z_i e \psi_0 / k_B T| \ll 1$ for all z_i. The result is then given by (4.62). If, in addition, the film thickness is much larger than the Debye length, (4.64) can be used. For symmetric electrolytes, the electrostatic disjoining pressure for arbitrary wall potentials can be found from (4.51) as long as h/λ_D is large.

4.3.4   Application to Film Drainage

Let us revisit the problem of drainage of a liquid film formed between a flattened gas bubble and a solid wall, as discussed in Chap. 3. In experiments conducted with aqueous solutions of NaCl [65], the disjoining pressure is usually dominated by its electrostatic component, so we neglect the effect of the London–van der Waals forces in the present section. Since the liquid film thickness in experiments

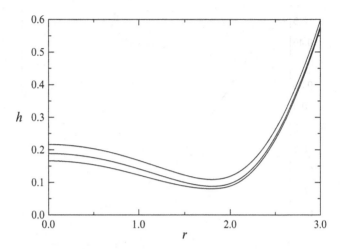

Fig. 4.8 Interface shapes at equal time intervals, $\Delta t = 50$, for film drainage governed by the electrostatic component of disjoining pressure, $\beta = 10^3$, $\chi = 10^2$

is typically much larger than the Debye length and the liquid is a symmetric electrolyte, the *nondimensional* pressure boundary condition at the liquid–gas interface can now be written as:

$$p = h_{rr} + r^{-1}h_r + \beta e^{-\chi h} \quad \text{at} \quad z = h(r,t), \tag{4.65}$$

where the first two terms on the right-hand side represent the capillary pressure jump (expressed in terms of the nondimensional variables defined in Chap. 3) and the third term comes from the scaled version of (4.51) so that

$$\beta = \frac{64n_0 k_B T}{\sigma/R_0} \tanh^2\left(\frac{ez\psi_0}{4k_B T}\right), \qquad \chi = \frac{2Ca^{2/3}R_0}{\lambda_D}, \tag{4.66}$$

where R_0 is the bubble radius away from the wall, σ is the surface tension at the gas–liquid interface, and Ca is the capillary number.

The evolution equation for the film thickness, $z = h(r,t)$, is obtained in the limit of small capillary numbers by following the procedure outlined in Chap. 3, except that (4.65) has to be used instead of (3.37) as the boundary condition for pressure. The resulting equation is

$$h_t + (3r)^{-1}\left[rh^3\left(h_{rr} + r^{-1}h_r + \beta e^{-\chi h}\right)_r\right]_r = 0. \tag{4.67}$$

The interface shapes obtained from the numerical solution of this equation with the conditions (3.42)–(3.45) are shown in Fig. 4.8. Formation of a dimple is seen as the liquid flows out of the narrow gap between the bubble and the wall. The interface

Fig. 4.9 Dimensional
equilibrium film thickness for
aqueous solution of NaCl on
quartz predicted by (4.68),
solid line, and measured
experimentally by Hewitt
et al. [65], filled squares

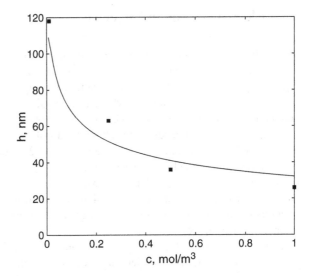

then flattens so that a static configuration is approached in which a flat film is in equilibrium with a curved meniscus due to the effects of disjoining pressure. The thickness of this film is found from the condition $\beta e^{-\chi h} = 2$ and can be expressed in *dimensional* terms as:

$$h^* = \frac{\lambda_D}{2}\ln\left[\frac{32n_0 k_B T}{\sigma/R_0}\tanh^2\left(\frac{ez\psi_0}{4k_B T}\right)\right]. \tag{4.68}$$

It is interesting to note that the interface shapes seen in Fig. 4.8 are not dramatically different from the ones predicted by the van der Waals model of disjoining pressure. Since the parameters of the electrostatic component of disjoining pressure are often easier to measure than the Hamaker constant, several detailed comparisons between theory and experiments have been made for both time scales and interface shapes in draining films, showing good agreement under a variety of different conditions [37, 91].

As an example of comparison between theory and experiment, consider the values of the equilibrium film thickness found from (4.68), shown by the solid line in Fig. 4.9 and the experimentally observed values for aqueous NaCl solution of different concentrations, shown by the squares. The dimensional electric potential at the water–quartz interface is obtained by a logarithmic fit of the experimental data quoted in Table 2 of [65]. The value of the bubble radius away from the wall is taken to be 1.09 mm. The agreement between the model and the experimental data in Fig. 4.9 is reasonable given the uncertainty in measurements of such small film thicknesses and also of the wall potential. The calculation of the capillary pressure jump in the meniscus region neglects the effects of gravity and the electric charges at the liquid–gas interface away from the wall; both of these can contribute to the discrepancy between theory and experiment.

4.4   Electrowetting on Dielectric

The wetting properties of a liquid on a solid substrate can be modified by an external electric field, a phenomenon knows as electrowetting. As an illustration, consider a liquid droplet on a flat solid substrate, sketched in Fig. 4.10. Voltage can be applied between two electrodes, one of which is typically a metal rod immersed in the conducting liquid as shown in the sketch, and the other one is a layer of metal under an insulating layer in the substrate. The static droplet shapes can be found using the equations of capillary statics as discussed in Sect. 1.2 except that the contact angle θ is now a function of the local electric field. In fact, in a typical experiment, the value of the contact angle rapidly decreases when the voltage is applied, resulting in the change of droplet shape shown in Fig. 4.10. For now, we assume that the droplet is in contact with air, although in experiments the medium above the liquid droplet is often another liquid, e.g., oil.

In the absence of electric fields, the contact angle θ is determined from the Young's equation, (1.4). The derivation of this equation presented in Sect. 1.2 is based on consideration of the local force balance near the contact line. Alternatively, the same equation can be derived from thermodynamic arguments as follows. For a static droplet on a solid substrate in the absence of electric fields, consider an arbitrary shape region which completely encloses the droplet; the boundary of this region is shown by the dashed line in the sketch in Fig. 4.11a and N_i moles of each phase ($i = s, l, g$) are inside the region at equilibrium. When small departures from equilibrium are considered, the values of N_i are assumed to remain constant, so that the matter initially inside the region forms a closed thermodynamic system (as defined in Appendix A). We seek equilibrium of the system at constant temperature and volume, which means that the Helmholtz free energy F has to be minimized (see Appendix A for general discussion of conditions of thermodynamic equilibrium). The contribution of interfaces to the overall Helmholtz free energy can be written as:

$$\tilde{F} = A_{lg}\sigma + A_{sl}\sigma_{sl} + A_{sg}\sigma_{sg}. \tag{4.69}$$

Here, A_{lg} is the area of the droplet surface, A_{sl} is the area of the solid–liquid interface under the droplet, A_{sg} is the area of the portion of the solid–gas interface inside the closed thermodynamic system considered; σ_{sl} and σ_{sg} denote surface tensions of the solid–liquid and solid–gas interfaces, respectively. We use the

Fig. 4.10 Electrowetting on dielectric: contact angle is decreased when the voltage is on (sketch on the right) compared to the value for zero voltage (sketch on the left)

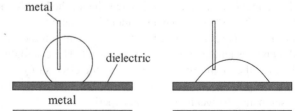

Fig. 4.11 Sketches illustrating thermodynamic system considered in the derivation of the classical Young's equation (**a**) and the geometric construction used in the derivation (**b**)

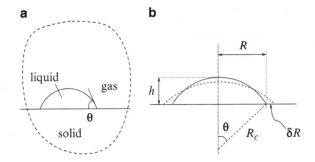

thermodynamic definition of the surface tension as the derivative of the Helmholtz free energy with respect to the interfacial area taken under the conditions of constant temperature, volumes, and the values of N_i in the three phases. In most situations of practical importance, this definition turns out to be equivalent to the one introduced in Sect. 1.1, although it is important to note that the two definitions are not identical [122].

Suppose the droplet in Fig. 4.11a is axisymmetric and the effect of gravity is negligible, so that the gas–liquid interface is part of a sphere of radius denoted by R_c. Using simple geometric considerations, the value of R_c can be expressed in terms of the radius of the wetted part of the substrate R and the droplet height h according to

$$R_c = \frac{R^2 + h^2}{2h}. \tag{4.70}$$

The condition of minimum Helmholtz free energy implies that the variation of $\tilde{F}$ is zero when the value of R is varied by a small amount δR while keeping the droplet volume,

$$V_d = \frac{\pi h}{2}\left(R^2 + \frac{1}{3}h^2\right), \tag{4.71}$$

constant. The dashed line in Fig. 4.11b shows corresponding droplet shape. The condition of zero variation of the droplet volume δV_d together with (4.70), gives a relationship between δh and δR in the form

$$\delta h = -\frac{R}{R_c}\delta R. \tag{4.72}$$

The variation of the area of the liquid–gas interface, $A_{lg} = \pi(R^2 + h^2)$, can be written as $\delta A_{lg} = 2\pi(R\delta R + h\delta h)$ or, using (4.72), as:

$$\delta A_{lg} = 2\pi R\delta R\frac{R_c - h}{R_c} = 2\pi R\cos\theta\,\delta R. \tag{4.73}$$

Fig. 4.12 A sketch illustrating calculation of the energy of the electric field in the layer of dielectric under the droplet when a voltage is applied

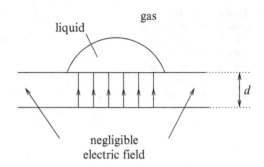

The variations of solid–liquid and solid–gas areas are expressed in terms of the area between two circles, of the radii $R + \delta R$ and R,

$$\delta A_{sl} = -\delta A_{sg} = 2\pi R \delta R. \tag{4.74}$$

The condition of $\delta \tilde{F} = 0$ together with (4.73) and (4.74) leads to the classical Young's equation,

$$\sigma \cos \theta = \sigma_{sg} - \sigma_{sl}. \tag{4.75}$$

In many experimental situations, the conditions of constant pressure and temperature are more relevant than those of constant volume and temperature and the surface tension is then defined as the derivative of the Gibbs free energy with respect to the surface area for constant temperature (T), pressures (P_i), and values of N_i in all phases. In fact, this definition of surface tension is considered standard in chemistry and chemical engineering literatures. If one assumes the conditions of constant P_i, T, and N_i to be satisfied for the closed thermodynamic system shown in Fig. 4.11a and proceeds with minimization of the Gibbs free energy of the system, the same Young's equation for the contact angle is obtained.

When a voltage V is applied in the electrowetting configuration of Fig. 4.10, the electric field is localized in the portion of the dielectric layer directly under the droplet, as shown in Fig. 4.12. We assume that both the liquid and the layer of metal under the dielectric are perfect conductors. In the part of the dielectric under the dry area, the electric potential ψ satisfies the Laplace's equation with the conditions of given potential at the bottom and zero derivative at the top of the layer. The only solution of this system is $\psi = \text{const.}$, so the electric field there is zero. There is a transition zone near the edge of the droplet in which the electric field in the dielectric changes from zero away from the droplet to V/d under it, where d is the dielectric layer thickness. Since the value of d it typically much smaller than the wetted radius of the substrate, R, the details of the transition zone do not have a significant effect on the solution and we can assume that there is a step-like change in electric field in the dielectric at the location corresponding to the edge of the droplet. Thus, the area of the base of the cylindrical region in which the electric field is nonzero is equal to A_{sl} and the electric field is uniform there. The electrostatic energy in this region can be calculated from the formula for energy storage in a simple capacitor which consists of two flat conducting plates and a layer of dielectric between them, $\frac{1}{2}C_D V^2$, where the capacitance C_D is

$$C_D = \frac{\varepsilon_D A_{sl}}{d} \tag{4.76}$$

and ε_D is the dielectric permittivity. Assuming that the entire region of nonzero electric field is inside the closed thermodynamic system shown in Fig. 4.11a and taking into account the electrostatic contribution to the energy of the system, (4.69) now reads [95]

$$\tilde{F} = A_{lg}\sigma + A_{sl}\left(\sigma_{sl} - \frac{\varepsilon_D V^2}{2d}\right) + A_{sg}\sigma_{sg}. \tag{4.77}$$

Minimization of the free energy leads to the formula

$$\cos\theta = \cos\theta_0 + \frac{\varepsilon_D V^2}{2d\sigma}, \tag{4.78}$$

which relates the contact angle θ to the value of the applied voltage V and the contact angle θ_0 at zero voltage, found from the classical Young's equation. The expression for the contact angle in terms of voltage given by (4.78), sometimes referred to as the Lippmann–Young equation, turns out to be in good agreement with experiments, as discussed, e.g., in [95].

Studies of electrowetting resulted in development of novel display technologies. For example, Hayes and Feenstra [63] demonstrated that a thin film of colored oil confined between four side walls on a $500 \times 500\,\mu m^2$ solid surface can be pushed into a corner by applying a voltage. The phenomenon is similar to the one sketched in Fig. 4.10 except that the top of the oil droplet is in contact with water rather than air, and the droplet height is increased rather than decreased when the voltage is applied, resulting in a transition from a nearly flat film of oil to a small droplet in a corner. The technological significance of this configuration is that it can be used as a pixel in a display image. By applying local voltage, the pixel color can be changed quickly from the color of the oil (when it covers the substrate) to the color of the substrate itself (usually white) when the oil droplet is localized in the corner. For sufficiently high voltage, the corner droplet can be made small enough so that it does not obscure the reflected light from the white substrate.

Other potential applications of electrowetting include liquid lenses with tunable focal length and manipulation of liquid droplets in microfluidic devices. The latter includes both open systems (droplets on substrates) and confined droplets. The basic principle is illustrated in Fig. 4.13. The contact angle on the right, controlled by the right electrode, can be changed independently from the contact angle on the left, resulting in a force which moves the droplet through the channel. This principle is the basis of many devices used in the so-called digital microfluidics [16].

Fig. 4.13 A sketch
illustrating application of
electrowetting to transport of
droplets in microchannels

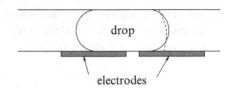

4.5   Notes on Literature

Introduction to electrokinetic phenomena can be found in Probstein [102], Chap.
5 of Adamson and Gast [2], and Levich [88]. More recent developments in the
field are covered, e.g., in Andelman [8], Butt and Kappl [28], Chang and Yeo [33],
Kirby [77], and Li [89]. Electrowetting and its applications are discussed in a review
article of Mugele and Baret [95] and two recently published books [16, 64].

Chapter 5
Flows and Interface Shapes Near Structured Surfaces

5.1 Introduction

It has long been recognized that properties of solid surfaces used in fluid mechanics experiments can have a significant effect on fluid flow, especially when contact lines are present. For many decades, commonly used approaches to controlling surface properties involved changing either their chemical composition or the degree of roughness (e.g., by polishing the surface to make is smoother). The actual geometric profiles of the solid surfaces at the nano-scale remained essentially random, with the only controllable parameter being the average degree of roughness. However, recent remarkable advances in microfabrication, as discussed, e.g., in the review article of Quéré [104], resulted in the ability to fabricate surfaces with well-defined small-scale structures. These surfaces are referred to as micro- or nano-structured; they include a wide variety of shapes, e.g., arrays of pillars of circular (Fig. 5.1a) or rectangular (Fig. 5.1b) cross-sections, rectangular grooves (Fig. 5.1c), as well as arrays of bumps and cavities. The sizes of surface features such as grooves, bumps, or cavities can be as small as 10 nm. For the geometry shown in Fig. 5.1a, pillar heights are typically of the order of 10 μm, while the cross-sectional dimensions are near 1 μm.

The small-scale structure can have a significant effect on the macroscopic fluid flow phenomena such as behavior of liquid droplets on the structured substrate. This is illustrated by a significant increase in the contact angle on surfaces of the type shown in Fig. 5.1a compared to the value on a flat smooth surface made of the same material. Here and below, we assume that the contact angle is determined based on macroscopic observations. When the contact angle is near 160° or above, the surface is commonly referred to as "superhydrophobic" [104]. The term "ultrahydrophobic" is used for the case when the contact angle is equal to 180°, although some authors use it to describe any superhydrophobic surface.

Experimental studies of superhydrophobic structured surfaces reveal a number of interesting phenomena [104]. For example, liquid droplets tend to roll rather

V.S. Ajaev, *Interfacial Fluid Mechanics: A Mathematical Modeling Approach*,
DOI 10.1007/978-1-4614-1341-7_5, © Springer Science+Business Media, LLC 2012

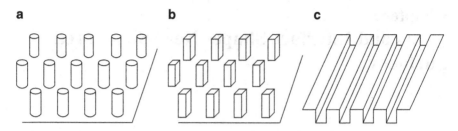

Fig. 5.1 Examples of structured surfaces: arrays of pillars of circular (**a**) and rectangular (**b**) cross-sections; (**c**) rectangular grooves

than slide when moving down an inclined superhydrophobic plate under the action of gravity, and even a very small inclination angle is sufficient to set them in motion. This is in contrast to nonstructured inclined surfaces on which droplets often remain stuck even for large inclination angles. Furthermore, when a drop impacts a horizontal superhydrophobic surface, it can bounce back almost as if it were made of a solid material. While deformed during the impact, the drop shows no tendency to spread over the surface and does not seem to lose a significant amount of its kinetic energy [107].

It is interesting to note that superhydrophobic surfaces can be found in nature. Notable examples include lotus leaves, butterfly wings, and the *Stenocara* beetle which has superhydrophobic stripes used to transport condensed water droplets to its mouth [98]. The latter mechanism is essential for survival in the desert areas where morning dew is an important source of otherwise scarce water. A recent study of compound mosquito eyes showed that they remain dry even when exposed to tiny droplets of water [54]. This mechanism allows the insect to maintain its vision while flying through fog.

Superhydrophobic surfaces have potential applications as the so-called self-cleaning surfaces [19]. A well-known mechanism of contamination of surfaces used in everyday life is due to the so-called "coffee-ring phenomenon," described as follows. Consider a liquid droplet which contains small solid particles, placed on a flat solid surface. When the droplet evaporates, the particles remain on the solid surface and form a stain. On superhydrophobic surfaces small droplets are not expected to get stuck. Instead, they will roll off the surface taking the solid particles with them. Thus, contamination of such surfaces is drastically reduced and they are often referred to as self-cleaning (although this term is arguably misleading). The lotus leaf, considered to be a symbol of purity in many human cultures, is an example of a system with self-cleaning properties encountered in nature.

Another important property of structured surfaces is that viscous drag for fluid flow past such surface can be significantly smaller than for a flat solid surface made of the same material. The origin of this phenomenon is now well-understood and discussed in detail below in Sect. 5.3, while its potential applications in microfluidics include drag reduction for viscous flows in microchannels, resulting in increased liquid transport rates.

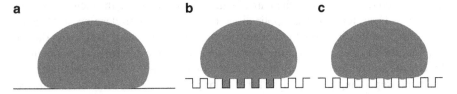

Fig. 5.2 Static droplet shapes on different surfaces: (**a**) classical case of smooth flat surface, (**b**) Wenzel configuration, (**c**) Cassie–Baxter configuration

5.2   Static Droplets: Wenzel and Cassie–Baxter Configurations

To illustrate the effects of surface structuring on the shapes of fluid interfaces in contact with solid surfaces we consider a simple problem of finding the static shape of a liquid droplet on a solid substrate. For the idealized case of a smooth flat surface, illustrated in Fig. 5.2a, the droplet shape is found by solving the equations of capillary statics together with the condition of a fixed contact angle θ between the liquid–gas interface and the solid. The contact angle can be expressed in terms of surface tension values at the three interfaces (liquid–gas, σ_{lg}, solid–gas, σ_{sg}, and solid–liquid, σ_{sl}) according to Young's equation:

$$\sigma_{lg} \cos \theta = \sigma_{sg} - \sigma_{sl}. \tag{5.1}$$

This equation was derived in Sect. 1.2 using a local force balance near the contact line. An alternative derivation based on thermodynamic arguments was provided in Sect. 4.4.

Wenzel [137] pointed out in his well-known study that on rough surfaces Young's equation has to be modified to read

$$\sigma_{lg} \cos \theta^{W} = r(\sigma_{sg} - \sigma_{sl}), \tag{5.2}$$

where θ^{W} is the contact angle found from macroscale observations, r is the roughness factor, i.e., the ratio of the actual solid surface area to the area of the flat surface placed at the average location of the rough surface. Equation 5.2 reflects the fact that the energies of the solid–gas and solid–liquid interfaces are proportional to the actual surface area of the rough solid. In reference to droplets on structured surfaces, the Wenzel model describes a situation shown in Fig. 5.2b, with the grooves (or spaces between the pillars) under the droplet completely filled with liquid. However, for structured surfaces a different, so-called Cassie–Baxter configuration, is also possible such that liquid does not enter the grooves/spaces between pillars and air is trapped there, as sketched in Fig. 5.2c. Many important phenomena discussed in Sect. 5.1, e.g., bouncing of droplets on superhydrophobic surfaces and significant drag reduction in channels with structured walls, have been observed in situations when air remains trapped between the elements of the structure. Thus, for practical applications ranging from self-cleaning surfaces to microfluidics, it is important to understand the conditions when the Cassie–Baxter configuration is preferred over the Wenzel state. A simple criterion can be obtained by comparing the free energies of the two configurations [105]. The fluid interface segments separating the liquid and the air trapped inside the structure should be nearly flat since the capillary pressure jump at these segments should be the same as at the drop surface away from the wall (except for a small correction due to hydrostatic pressure differences). Thus, the bottom of the drop is nearly flat, with a fraction of it denoted by ϕ_s being solid–liquid and the rest liquid–gas segments. The change in the free energy per unit area of the base of the drop between the Wenzel and the Cassie–Baxter configurations is then

$$(r - \phi_s)(\sigma_{sl} - \sigma_{sg}) - (1 - \phi_s)\sigma_{lg}. \tag{5.3}$$

If it is positive, the liquid will enter the grooves so that the energy of the system is reduced. This defines the critical angle condition

$$\cos \theta^c = -\frac{1 - \phi_s}{r - \phi_s}. \tag{5.4}$$

The value of the apparent contact angle for the Cassie–Baxter configuration can be estimated by considering the energy change when the contact line advances by a small amount so that the solid–gas interface segments at the top of the structure (e.g., the tops of the cylindrical pillars) are replaced with solid–liquid segments of the same area and additional liquid–gas interfaces are created in the form of the menisci between the elements of the structure [31]. Therefore,

$$\sigma_{lg} \cos \theta^{CB} = \phi_s(\sigma_{sg} - \sigma_{sl}) + \sigma_{lg}(\phi_s - 1). \tag{5.5}$$

This formula is applied above the critical angle. Thus, based on our simple energy argument, the apparent contact angle θ^* is given by (5.2) below the critical angle and by (5.5) above it. Combining these equations and the classical Young's equation (5.1) allows one to relate the apparent contact angle on the structured surface θ^* to the contact angle θ on a flat surface made of the same material by using

$$\cos \theta^* = \begin{cases} r\cos\theta, & \theta \leq \theta^c, \\ -1 + \phi_s(1 + \cos\theta), & \theta^c < \theta. \end{cases} \tag{5.6}$$

This result is represented graphically by the solid lines in the sketch in Fig. 5.3.

In reality, the situation is often more complicated than suggested by the simple energy argument used above [18, 82]. For example, the metastable states of the Cassie–Baxter type, corresponding to the dashed line in Fig. 5.3, have been observed and sharp transitions from these states to the Wenzel-type behavior has been recorded in experiments. The details of such transitions are still a subject of ongoing

Fig. 5.3 The predictions of the apparent contact angle on a structured surface based on Cassie–Baxter and Wenzel models (*solid lines*). The dashed line corresponds to metastable states of the Cassie–Baxter type. After [82]

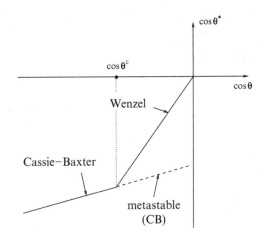

research, with some results surveyed in the article of Quéré [105]. It is important to emphasize that the arguments of the present section only apply to solid surfaces with regular periodic structure and the large ratio of drop size to the scale of the structure, so they are likely to fail for solid surfaces with isolated defects and for sufficiently small droplets.

Finally, we note that while the terms "Wenzel" and "Cassie–Baxter" are introduced here in the context of droplets on structured surfaces, they are used to describe any geometric configurations in which liquid penetrates or does not penetrate into the structure, respectively.

5.3 Slip Flow in a Channel

5.3.1 Series Solution for Transverse Grooves

Viscous flow in a channel with flat walls under the action of an imposed pressure gradient has been discussed in Sect. 1.3. Let us now investigate the effects of wall structuring on such flow. Consider a channel with walls structured by a periodic array of grooves aligned in the direction of the z-axis [129]. The pressure gradient G is in the x-direction and the model is two-dimensional so that velocity and pressure fields are assumed to be functions of x and y only. The grooves are filled with gas phase as sketched in Fig. 5.4 (a Cassie–Baxter type configuration) and the interfaces separating gas in the grooves from the liquid are assumed flat. Thus, the boundaries of the liquid domain are flat and parallel to each other; the half-distance between them, d, is used as the scale for all length variables. The groove pattern is periodic with the scaled periodicity length L. Note that Fig. 5.4 shows only a section of the channel corresponding to $-L/2 \leq x \leq L/2$. The scaled groove width is denoted

Fig. 5.4 Geometric
configuration of a channel
with gas-filled grooves
transverse to the flow

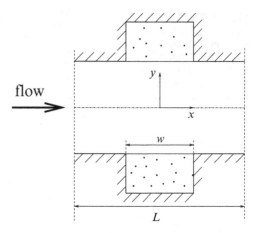

by w. The ratio $\delta = w/L$ is an important parameter determining the fraction of
the channel wall which is a liquid–gas rather than liquid–solid interface (note that
$\delta = 1 - \phi_s$).

Using the vorticity–stream function formulation for the Stokes flow, introduced
in Sect. 2.1.6, we write

$$\nabla^2 \omega = 0, \tag{5.7}$$

$$\nabla^2 \psi = -\omega. \tag{5.8}$$

The scales here are different from the ones used in Chap. 2: the vorticity ω is scaled
by Gd/μ and the stream function ψ by Gd^3/μ. The negative sign on the right-hand
side of (5.8) corresponds to the conventional definition of vorticity as curl of the
velocity field, $\omega = v_x - u_y$. The velocity components u and v, both scaled by Gd^2/μ,
can be expressed in terms of the stream function using the standard formulas,

$$u = \psi_y, \quad v = -\psi_x. \tag{5.9}$$

The solution is symmetric with respect to the line $y = 0$, so it is sufficient to consider
the upper half of the channel, $0 \leq y \leq 1$, and impose the symmetry conditions of
$v = 0$ and $u_y = 0$ at $y = 0$. These symmetry conditions can be written in terms of the
vorticity and the stream function (defined up to an additive constant) as:

$$\psi = 0 \quad \text{at} \quad y = 0, \tag{5.10}$$

$$\omega = 0 \quad \text{at} \quad y = 0. \tag{5.11}$$

The second of these conditions simply reflects the fact that there is no rotation of
fluid elements located on the symmetry line and can be formally derived by noticing
that $v = 0$ implies $v_x = 0$ and therefore $\omega = v_x - u_y = 0$ along the line $y = 0$.

The upper boundary of the domain ($y = 1$) consists of two parts: the solid–liquid interface at which the no-slip condition ($u = 0$) is satisfied, and the liquid–gas interface. At the latter we assume zero shear stress, $u_y = 0$, although for very small groove sizes the effect of gas viscosity is not negligible and therefore the Navier slip condition discussed in Sect. 1.5* may be more appropriate [15]. At both liquid–gas and solid–liquid interface segments, the vertical velocity component v is zero so $v_x = 0$ and the vorticity is proportional to the tangential stress at these interfaces. Using these observations and (5.9), the boundary conditions at $y = 1$ for the channel segment shown in Fig. 5.4 can be summarized as follows,

$$\omega = 0, \quad \psi = \psi_1 \quad \text{at} \quad |x| \le \frac{w}{2}, \ y = 1, \tag{5.12}$$

$$\psi_y = 0, \quad \psi = \psi_1 \quad \text{at} \quad \frac{w}{2} < |x| \le \frac{L}{2}, \ y = 1, \tag{5.13}$$

where ψ_1 is a constant. Even though ψ is defined up to an additive constant, we cannot set ψ_1 to zero here since the constant value of ψ at the symmetry line (set to zero in 5.10) is in general not the same as the constant value of ψ at $y = 1$.

Since we are interested in describing viscous flow in a long channel with a large number of grooves, the periodic conditions for both ψ and ω will be used at the side boundaries ($x = \pm L/2$). Furthermore, based on the shape and location of the groove, we look for solutions which are symmetric with respect to the y-axis.

It is convenient to write the solution of the flow problem as a superposition,

$$\psi = \psi_P + \tilde{\psi}, \qquad \omega = \omega_P + \tilde{\omega}, \tag{5.14}$$

where the subscript "P" corresponds to the flow in a channel with flat solid walls (no grooves), i.e., the classical two-dimensional Poiseuille flow discussed in Sect. 1.3. Based on the dimensional result obtained there, (1.24), the nondimensional vorticity and stream function for the Poiseuille flow can be written as:

$$\omega_P = y, \qquad \psi_P = \frac{y}{2} - \frac{y^3}{6}. \tag{5.15}$$

The correction to the vorticity due to the effect of the structuring satisfies the Laplace's equation

$$\nabla^2 \tilde{\omega} = 0. \tag{5.16}$$

Since the boundary conditions at $x = \pm L/2$ are periodic and $\tilde{\omega}(x) = \tilde{\omega}(-x)$, the solution for $\tilde{\omega}$ can be expressed as a Fourier cosine series,

$$\tilde{\omega} = Ay + B + \sum_{n=1}^{\infty} f_n(y)\cos(k_n x), \qquad k_n = \frac{2\pi n}{L}. \tag{5.17}$$

A and B are constants. Based on (5.16), the functions $f_n(y)$ have to satisfy the ordinary differential equations

$$f_n'' - k_n^2 f_n = 0, \qquad n = 1, 2, \ldots,$$

which have the general solutions

$$f_n = a_n \cosh(k_n y) + b_n \sinh(k_n y).$$

Using the boundary condition (5.11), we immediately obtain $B = 0$ and $a_n = 0$. Furthermore, since the average pressure gradient is equal to unity in our scaled variables, the coefficient A has to be set to zero. Thus,

$$\tilde{\omega} = \sum_{n=1}^{\infty} b_n \sinh(k_n y) \cos(k_n x). \tag{5.18}$$

Substituting this result into the equation for the correction to the stream function,

$$\nabla^2 \tilde{\psi} = -\tilde{\omega}, \tag{5.19}$$

we obtain

$$g_n'' - k_n^2 g_n = -b_n \sinh(k_n y), \qquad n = 1, 2, \ldots, \tag{5.20}$$

where the functions $g_n(y)$ are the coefficients of the Fourier series for the stream function correction,

$$\tilde{\psi} = Cy + D + \sum_{n=1}^{\infty} g_n(y) \cos(k_n x). \tag{5.21}$$

The general solutions of (5.20) are

$$g_n = c_n \cosh(k_n y) + d_n \sinh(k_n y) - \frac{b_n y}{2 k_n} \cosh(k_n y). \tag{5.22}$$

By applying the boundary condition (5.10), we find $D = 0$ and $c_n = 0$. According to (5.12)–(5.13), $\tilde{\psi}$ is constant everywhere along the line $y = 1$, so $g_n(1) = 0$ and therefore

$$b_n = 2 d_n k_n \tanh k_n. \tag{5.23}$$

The formulas for $\tilde{\omega}$ and $\tilde{\psi}$ can now be written in the form

$$\tilde{\omega} = \sum_{n=1}^{\infty} 2 d_n k_n \tanh k_n \sinh(k_n y) \cos(k_n x), \tag{5.24}$$

$$\tilde{\psi} = Cy + \sum_{n=1}^{\infty} d_n \left(\sinh(k_n y) - y \tanh k_n \cosh(k_n y) \right) \cos(k_n x). \tag{5.25}$$

Using the conditions (5.12)–(5.13) together with (5.15), the following system of equations for the unknown coefficients d_n and C is obtained,

$$\frac{1}{2} + \sum_{n=1}^{\infty} d_n \alpha_n \cos(k_n x) = 0, \qquad |x| \le \frac{w}{2}, \tag{5.26}$$

$$C + \sum_{n=1}^{\infty} d_n \beta_n \cos(k_n x) = 0, \qquad \frac{w}{2} < |x| \le \frac{L}{2}, \tag{5.27}$$

where to simplify the notation we introduced

$$\alpha_n = k_n \tanh k_n \sinh k_n, \qquad \beta_n = \frac{k_n}{\cosh k_n} - \sinh k_n. \tag{5.28}$$

Since the infinite series in (5.26) and (5.27) are both convergent, they can be approximated by their truncated versions, with summation over n from 1 to $N-1$ instead of infinity, where N is a sufficiently large integer number. Multiplying the truncated version of (5.26) by $\cos(k_m x)$, where m is an integer between 0 and $N-1$, and integrating in x, we obtain

$$\frac{1}{2} \int_0^{w/2} \cos(k_m x)\, dx + \sum_{n=1}^{N-1} d_n \alpha_n \int_0^{w/2} \cos(k_n x)\cos(k_m x)\, dx = 0. \tag{5.29}$$

Applying the same procedure to (5.27) results in

$$C \int_{w/2}^{L/2} \cos(k_m x)\, dx + \sum_{n=1}^{N-1} d_n \beta_n \int_{w/2}^{L/2} \cos(k_n x)\cos(k_m x)\, dx = 0. \tag{5.30}$$

Adding (5.29) and (5.30) together for each m and evaluating the integrals on the left-hand side results in a linear system of equations for the coefficients C and d_n. It is convenient to write this system in the matrix–vector form, as $A\mathbf{g} = \mathbf{r}$, where the vector of the yet unknown coefficients $\mathbf{g}$ has the components $g_1 = C$, $g_i = d_{i-1}$ ($i = 2,\ldots N$), the right-hand side vector $\mathbf{r}$ is defined in terms of $\delta \equiv w/L$ by $r_1 = -\pi\delta$, $r_i = -(i-1)^{-1}\sin(\pi\delta(i-1))$ for $i > 1$, and the entries of the matrix A are given by:

$$A_{ij} = (\alpha_{j-1} - \beta_{j-1})\left[\frac{\sin(\pi\delta(i+j-2))}{i+j-2} + \frac{\sin(\pi\delta(i-j))}{i-j}\right], \qquad i \ne j, \tag{5.31}$$

$$A_{jj} = \pi\left(\alpha_{j-1}\delta + \beta_{j-1}(1-\delta)\right) + \frac{\alpha_{j-1} - \beta_{j-1}}{2(j-1)}\sin(2\pi\delta(j-1)), \tag{5.32}$$

for $j > 1$ and $A_{11} = 2\pi(1-\delta)$, $A_{i1} = -2(i-1)^{-1}\sin(\pi\delta(i-1))$ ($i = 2,\ldots N$).

It may seem that finding the vector **g** is now completely straightforward and can be accomplished by using any standard linear system solver. However, it turns out that typical solvers fail for this problem as N is increased above a certain value, typically of the order of 10 (e.g., MATLAB fails to provide a solution of the problem for $\delta = 0.5, L = 0.1, N = 20$). The numerical difficulties encountered by even the most robust and widely used linear solvers are due to the extremely fast increase of the absolute values of α_n and β_n as n is increased. Indeed, according to (5.28) the asymptotic behavior of these coefficients in the limit of $n \to \infty$ is

$$\alpha_n \sim \frac{1}{2} k_n e^{k_n}, \quad \beta_n \sim -\frac{1}{2} e^{k_n}. \tag{5.33}$$

This implies that some entries of the matrix A are very large, which in turn results in the matrix being ill-conditioned and the numerical procedure for finding the inverse being inaccurate. Fortunately, the numerical difficulties can be avoided if instead of the rapidly growing coefficients α_n and β_n one uses their rescaled versions defined by:

$$\hat{\alpha}_n = e^{-k_n} \alpha_n, \quad \hat{\beta}_n = e^{-k_n} \beta_n. \tag{5.34}$$

A matrix $\hat{A}$ is defined using the same formulas as the matrix A above except that the coefficients α_{j-1} and β_{j-1} are replaced by $\hat{\alpha}_{j-1}$ and $\hat{\beta}_{j-1}$. The rescaled solution vector $\hat{\mathbf{g}}$ is introduced such that its components are

$$\hat{g}_i = g_i e^{-k_{i-1}}, \quad i = 1, 2 \ldots N. \tag{5.35}$$

The problem is then reduced to solving the system $\hat{A}\hat{\mathbf{g}} = \mathbf{r}$, from which the vector $\hat{\mathbf{g}}$ is found without any numerical difficulties using standard linear solvers, as illustrated by the MATLAB code in the Appendix, Sect. B.5. Using (5.35), the constants appearing in (5.25) can now be expressed as:

$$C = \hat{g}_1, \quad d_{i-1} = \hat{g}_i e^{k_{i-1}}, \quad i = 2 \ldots N. \tag{5.36}$$

The perturbation stream function $\tilde{\psi}$ is then found by using the truncated version of the series from (5.25). Note that as a result of the structuring, the average flow velocity in the direction along the channel is changed by a nonzero amount, determined by the value of the coefficient C. Thus, the transport rate is increased due to the presence of the gas-filled grooves, which can be explained by liquid slippage at the liquid–gas interfaces in the grooves. The x-dependent part of the solution for $\tilde{\psi}$ determines the perturbation flow pattern not contributing to the average transport rate: typical streamlines of this recirculation pattern obtained from the MATLAB code of Sect. B.5 are shown in Fig. 5.5.

To summarize, the flow velocity in the channel is a superposition of the classical parabolic velocity profile for the channel with flat walls, the constant correction in the direction along the channel equal to the value of $C = \hat{g}_1$, and the recirculation pattern illustrated in Fig. 5.5.

Fig. 5.5 Streamlines
corresponding to the
perturbation stream function
for $L = 2$, $\delta = 0.5$ with the
average velocity solution in
the x-direction subtracted out

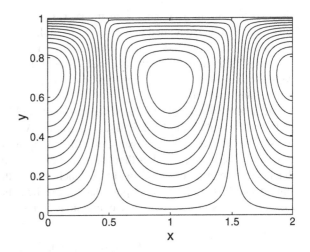

5.3.2   Effective Slip Length

In practical applications, the number of grooves is typically very large, so it is
natural to try to describe their effect in terms of effective properties of the structured
surface. Since the Cassie–Baxter configuration implies that parts of the lower
boundary of the liquid domain have perfect slip (the liquid–gas interface segments
in the grooves), it can be expected that the boundary condition of no-slip should be
replaced by some effective boundary condition. Following Lauga and Stone [86], we
define an *effective* slip length as the slip length of the parallel flow which is driven
by the same pressure gradient and has the same flow rate as the flow of interest.
Such parallel slip flow in a channel with flat (nonstructured) walls was discussed in
Sect. 1.5*. The nondimensional version of the velocity profile obtained there (and
given by 1.92) is

$$u = \frac{1}{2}(1 - y^2) + \lambda, \tag{5.37}$$

where u is the velocity scaled by Gd^2/μ and λ is the slip length scaled by d. The
pressure gradient G is constant along the channel. The corresponding scaled flow
rate (obtained by integrating u across the entire channel width) can be written as:

$$Q = 2\left(\frac{1}{3} + \lambda\right). \tag{5.38}$$

This formula can now be applied to express the effective slip length λ_{eff} for the flow
in a channel with structured walls in terms of the flow rate as:

$$\lambda_{\text{eff}} = \frac{Q}{2} - \frac{1}{3}. \tag{5.39}$$

The rate of flow in a channel with transverse grooves discussed in the previous subsection can be calculated based on the formulas for the stream function, (5.15) and (5.25). The x-dependent part of the expansion in (5.25) does not contribute to the average flow rate along the channel, so only the coefficient C enters the final expression. In fact, after the contribution of the parabolic velocity profile is taken into account, we find $\lambda_{\text{eff}} = C$.

Lauga and Stone [86] identify several asymptotic limits allowing one to obtain analytical formulas for the effective slip length. As an illustration, we consider one such limit, based on the assumption that L approaches zero but δ is fixed. In dimensional terms this implies that the periodicity length in the direction along the channel (of the order of groove width) is much smaller that the channel height. The limit of small L (i.e., large k_n) implies that the coefficients α_n and β_n can be replaced with the corresponding asymptotic expressions at $n \to \infty$, leading to

$$1 + \sum_{n=1}^{\infty} k_n \hat{d}_n \cos(k_n x) = 0, \quad |x| \leq \frac{w}{2}, \tag{5.40}$$

$$\frac{1}{2}\hat{d}_0 + \sum_{n=1}^{\infty} \hat{d}_n \cos(k_n x) = 0, \quad \frac{w}{2} < |x| \leq \frac{L}{2}, \tag{5.41}$$

where $\hat{d}_0 = -4C$, $\hat{d}_{n-1} = \hat{g}_n$. The same system of equations can be written in terms of a new variable $z = (2\pi/L)|x|$ as:

$$\frac{L}{2\pi} + \sum_{n=1}^{\infty} n\hat{d}_n \cos nz = 0, \quad z \leq \pi\delta, \tag{5.42}$$

$$\frac{1}{2}\hat{d}_0 + \sum_{n=1}^{\infty} \hat{d}_n \cos nz = 0, \quad \pi\delta < z \leq \pi, \tag{5.43}$$

which is a special case of a problem considered in Sect. 5.4.3 of Sneddon [116]. Following the method introduced there, we observe that the coefficients $\hat{d}_n$ are the Fourier coefficients of a yet unknown function $g(z)$ such that it is identically zero on the interval $[\pi\delta, \pi]$. It is convenient to express this function in terms of an integral involving another unknown function $h(z)$ as:

$$g(z) = \cos\frac{z}{2} \int_z^{\pi\delta} \frac{h(t)\mathrm{d}t}{\sqrt{\cos z - \cos t}}. \tag{5.44}$$

Then from the standard formulas for the Fourier coefficients of $g(z)$, we obtain

$$\hat{d}_0 = \sqrt{2} \int_0^{\pi\delta} h(t)\mathrm{d}t, \tag{5.45}$$

$$\hat{d}_n = \frac{1}{\sqrt{2}} \int_0^{\pi\delta} h(t)[P_n(\cos t) + P_{n-1}(\cos t)]dt, \tag{5.46}$$

where $P_n(\cos\theta)$ are the Legendre polynomials and Mehlers integral representation has been used in writing the equations for $n > 0$ [116]. Integrating (5.42) results in the formula (applicable for z between 0 and $\pi\delta$),

$$\frac{L}{2\pi}z + \sum_{n=1}^\infty \hat{d}_n \sin nz = 0. \tag{5.47}$$

Substituting the above expressions for $\hat{d}_n$ into this formula leads to

$$\frac{L}{2\pi}z + \frac{1}{\sqrt{2}} \int_0^{\pi\delta} h(t) \sum_{n=1}^\infty [P_n(\cos t) + P_{n-1}(\cos t)] \sin nz\, dt = 0. \tag{5.48}$$

Using the summation formulas from Sect. 2.6 of [116], the result can be expressed as an integral equation for $h(z)$,

$$\frac{Lz}{2\pi} \sec\frac{z}{2} + \int_0^z \frac{h(t)dt}{\sqrt{\cos t - \cos z}} = 0, \tag{5.49}$$

which has the solution

$$h(z) = -\frac{L}{\pi^2}\frac{d}{dz} \int_0^z \frac{t\sin(t/2)dt}{\sqrt{\cos t - \cos z}}. \tag{5.50}$$

As discussed above, the effective slip length only depends on the zeroth coefficient in the Fourier expansion which is found from the solution for the function $h(z)$ by using (5.45). The resulting formula for the effective slip length is written in the form

$$\lambda_{\text{eff}} \sim \frac{L}{2\pi} \ln\left[\sec\left(\frac{\pi\delta}{2}\right)\right]. \tag{5.51}$$

This asymptotic result turns out to work remarkably well over a range of values of L, as illustrated by the plot in Fig. 5.6.

The formula for the effective slip length, (5.51), was derived based on the assumption that the flow direction is perpendicular to the direction of the grooves, used in all derivations described in the previous subsection. The case of grooves aligned in the flow direction can also be treated using Fourier expansions [100]. In the asymptotic limit of small L and finite δ the scaled effective slip length turns out to be twice the value given by (5.51). For arbitrary relative orientation of the

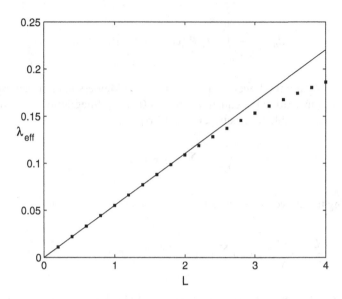

Fig. 5.6 The effective slip length for $\delta = 0.5$ as predicted by the series solution with $N = 2000$ (*filled squares*) and by the asymptotic formula (*solid line*) derived in the limit of $L \to 0$ and fixed δ

grooves and the flow direction, the slip length is characterized by the effective slip-length tensor [13, 52], which in our asymptotic limit can be expressed in terms of the orientation angle of the grooves and the value of λ_{eff} given by (5.51).

5.4* Electroosmotic Flow Over Structured Surface

In recent years, the presence of electric charges at structured surfaces and its effect on the flow near such surfaces received significant attention in the literature, as discussed e.g. in [10,121,133]. In the previous section it was observed that the transport rate for the pressure-driven flow in small channels is enchanced due to the presence of the grooves. It is natural to ask if structuring can have the same effect on the electroosmotic flow studied in Chap. 4. To answer this question, let us follow Bahga et al. [10] and consider the same Cassie-Baxter configuration as in the previous section but assume that the electric charges can be present at the interfaces. The charge densities are denoted by q_1 for solid-liquid and q_2 for liquid-gas interfaces and the external electric field E_0 is in the direction along the wall and perpendicular to the direction of the grooves. The channel height is assumed to be much larger than the periodicity length of the groove structure. The fluid flow velocity $\mathbf{u}$ is found by solving the Stokes flow model with the electrostatic body force,

$$\mu \Delta \mathbf{u} = -\varepsilon \Delta \psi \nabla \phi + \nabla p, \qquad (5.52)$$

$$\nabla \cdot \mathbf{u} = 0, \qquad (5.53)$$

where ε is the dielectric permittivity, p is pressure, ϕ is the total electrostatic potential in the liquid, ψ is its perturbation due to diffuse charge. The latter is found using the Debye-Hückel approximation, discussed in detail in Chap. 4,

$$\Delta \psi = \lambda_D^{-2} \psi, \tag{5.54}$$

where λ_D is the Debye length. The no-slip condition is applied at the solid-liquid interface segments, while perfect slip is assumed at the liquid-gas ones. Note that the system of governing equations is linear, so Fourier series expansions can be used to solve it, as was done for the case without charges in the previous section. Application of the boundary conditions leads to the dual series expansions. While these general formulas are rather complicated and not discussed here, several important physical results can be obtained using asymptotic methods by considering the limits of the thin ($\lambda_D \ll L$) and thick ($\lambda_D \gg L$) electrical double layers, where L now denotes the dimensional periodicity length of the groove structure. Let us first consider the case when only the solid-liquid interface segments are charged so that $q_1 = q_0$, $q_2 = 0$. The most important quantity for practical applications, the electroosmotic flow velocity u_{EO}, measured just outside the electrical double layer, for this case is described by the following formulas,

$$u_{EO} = -\frac{E_0 q_0 \lambda_D}{\mu}, \qquad \lambda_D \ll L, \tag{5.55}$$

$$u_{EO} = -\frac{(1-\delta)E_0 q_0 \lambda_D}{\mu}, \qquad \lambda_D \gg L. \tag{5.56}$$

The first of these asymptotic formulas clearly suggests that the flow enhancement is negligible for the case of thin electrical double layer, in agreement with the general result obtained by Squires [121]. For the thick electrical double layer, the velocity is even smaller than for $\lambda_D \ll L$. For the case of arbitrary L comparable to λ_D, no simple analytical formulas are available, but the flow velocity can be obtained numerically [10, 133]. Typical results are shown by the solid line in Fig. 5.7a for the case $\delta = 0.5$. Dashed line corresponds to the scaled electroosmotic flow velocity for the case when the grooves are oriented along the direction of the external electric field. The results clearly show that there is no significant flow enhancement over the entire range of values of the periodicity length L.

The situation can be dramatically different when electrical charges are present at the liquid-gas interface segments. As an illustration, we consider the special case of $q_1 = q_2 = q_0$. Remarkably, an exact solution can be obtained for this case for arbitrary double layer thickness in the form

$$u_{EO} = -\frac{E_0 q_0 \lambda_D}{\mu} \left(1 + \frac{L}{2\pi \lambda_D} \ln \left[\sec \frac{\pi \delta}{2} \right] \right). \tag{5.57}$$

This solution is shown for the case of $\delta = 0.5$ by the solid line in Fig. 5.7b. The flow velocity exceeds the value corresponding to the case when the charges are only

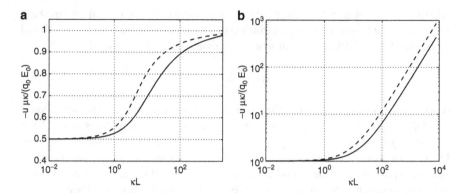

Fig. 5.7 The effective slip length for electroosmotic flow near charged structured surface with $q_1 = q_0, q_2 = 0$ (**a**) and $q_1 = q_2 = q_0$ (**b**); $\kappa = \lambda_D^{-1}$. From [133], reprinted with permission of IOP

present at the liquid-solid interface segments, by orders of magnitude. The same observation can be made for the case of longitudinal grooves, shown by the dashed line in Fig. 5.7b. Thus, the electroosmotic flow can be significantly enhanced by the slip effects when charges are present at the liquid-gas segments of the interface boundary.

We also note that the more general case of arbitrary orientation of the electric field with respect to the orientation of the grooves has been discussed in [10].

5.5 Notes on Literature

Wetting on structured surfaces is discussed in a review article of Quéré [105]. Detailed reviews of various flow phenomena related to structured, especially superhydrophobic surfaces can be found in [19, 104]. Flows in channels with structured walls are discussed in [86, 109, 129]. The key results on the electroosmotic flow near structured surfaces can be found in [10, 121, 133].

Chapter 6
Phase Change at Interfaces

6.1 Evaporation/Condensation at Flat Liquid–Vapor Interfaces

6.1.1 Thermodynamic Equilibrium Conditions

Consider a flat interface between a liquid and pure vapor of the same liquid. The equations of capillary statics discussed in Chap. 1 imply that if the effects of disjoining pressure are negligible, the interface is stationary when the pressures are equal to each other in the liquid and vapor phases. This can be understood as the condition of mechanical equilibrium: the normal forces acting on an interface segment from the liquid and the vapor sides have to balance each other. The condition of thermal equilibrium requires that the temperature is uniform. However, the conditions of uniform pressure and temperature are not sufficient to ensure equilibrium in a two-phase system. In addition, the chemical potentials of the liquid (μ_l) and vapor (μ_v) phases must be equal to each other so that there is no phase change, i.e., evaporation or condensation, at the vapor–liquid interface. In order to understand the origin of this additional condition, consider a simply connected region of arbitrary shape which contains both liquid and vapor. Suppose we consider processes during which the boundary of the region can deform, but the total quantity of matter inside, $N_l + N_v$, remains constant, making it a closed thermodynamic system as defined in Appendix A. Here, N_l and N_v refer to numbers of moles of liquid and vapor, respectively. The conditions of thermodynamic equilibrium at constant pressure and temperature imply that the Gibbs free energy of the system is minimized, so that $dG = 0$. Considering a phase change process at constant pressure and temperature, this condition can be written as:

$$\mu_l dN_l + \mu_v dN_v = 0. \tag{6.1}$$

V.S. Ajaev, *Interfacial Fluid Mechanics: A Mathematical Modeling Approach*,
DOI 10.1007/978-1-4614-1341-7_6, © Springer Science+Business Media, LLC 2012

Since $N_l + N_v$ is constant, we immediately obtain $\mu_l = \mu_v$. This condition implies that for a given pressure, equilibrium is only possible at a specific value of temperature, called the saturation temperature.

The points on the pressure–temperature diagram corresponding to equal values of μ_l and μ_v form a so-called vapor–liquid coexistence curve, $P(T)$. Since the chemical potentials can be interpreted as the molar Gibbs free energies of the phases, the standard formula for the differential of G [79, see also Appendix A], can be applied to $d\mu_k$ (k = v, l):

$$d\mu_k = -s_k dT + v_k dP, \tag{6.2}$$

where s_k and v_k are the molar entropies and volumes, respectively. Along the coexistence curve, $d\mu_v = d\mu_l$, so

$$\frac{dP}{dT} = \frac{s_v - s_l}{v_v - v_l}. \tag{6.3}$$

Assuming the molar volume of the liquid is small compared to that of the vapor and introducing the molar enthalpy of vaporization, Δh (also referred to as the latent heat of vaporization), we obtain

$$\frac{dP}{dT} = \frac{\Delta h}{T v_v}. \tag{6.4}$$

This formula, often referred to as the Clausius–Clapeyron equation, determines the slope of the vapor–liquid coexistence curve. Since the molar volume of an ideal gas is $v_v = RT/P$, the Clausius–Clapeyron equation can also be written as:

$$\frac{d\ln P}{dT} = \frac{\Delta h}{RT^2}. \tag{6.5}$$

If a specific latent heat (i.e., the latent heat per unit mass), $\mathscr{L}$, is used instead of the molar quantity Δh, (6.5) becomes

$$\frac{d\ln P}{dT} = \frac{\mathscr{L}M}{RT^2}, \tag{6.6}$$

where M is the molar mass.

The discussion of the equilibrium conditions so far has been based on the assumption that the effects of disjoining pressure, first introduced in Sect. 1.7, are negligible. For a liquid film on a solid surface, sketched in Fig. 6.1, the disjoining pressure can become important when the film thickness d is of the order of 100 nm or less, which is the case in a number of experiments. For example, films of pentane and heptane of thicknesses well below 100 nm have been observed on quartz substrates in careful experimental studies conducted by Wayner et al. [41, 76, 136]. For such ultrathin films, the chemical potential of the liquid is different from the bulk value of μ_l, resulting in a modified equilibrium condition,

$$\mu_l - v_l \Pi = \mu_v, \tag{6.7}$$

Fig. 6.1 Sketch of a uniform liquid film on a solid surface

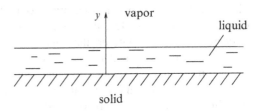

where Π is the disjoining pressure. In the absence of electric charges, the disjoining pressure can be approximated by:

$$\Pi = -\frac{A}{d^3}, \tag{6.8}$$

based on the theory of London–van der Waals forces as discussed in more detail in Sect. 1.7; A is the Hamaker constant.

Suppose the pressure in the vapor is P_v. Then, for thick films with negligible effects of disjoining pressure, the equilibrium will be reached when the liquid pressure is also equal to P_v, the temperature is uniform and equal to its saturation value T_{sat} corresponding to P_v, and

$$\mu_l(T_{sat}, P_v) = \mu_v(T_{sat}, P_v). \tag{6.9}$$

In the presence of the van der Waals disjoining pressure, the condition (6.7) implies that for the same vapor pressure equilibrium is reached at a temperature $T_{sat} + \Delta T$ such that

$$\mu_l(T_{sat} + \Delta T, P_v) + \frac{Av_l}{d^3} = \mu_v(T_{sat} + \Delta T, P_v). \tag{6.10}$$

Expanding both μ_l and μ_v in the Taylor series around $T = T_{sat}$ and observing that for $\Delta T \ll T_{sat}$ the terms which are nonlinear in ΔT can be neglected, we arrive at

$$\frac{Av_l}{d^3} = (s_l - s_v)\Delta T. \tag{6.11}$$

Here, we have also used the condition of equality of the chemical potentials of the bulk phases, (6.9). Since the liquid density ρ can be expressed as $\rho = M/v_l$ and the specific latent heat of evaporation is $\mathscr{L} = (s_v - s_l)T_{sat}/M$, (6.11) can be written in the form

$$\frac{A}{d^3} = -\frac{\rho \mathscr{L} \Delta T}{T_{sat}}. \tag{6.12}$$

This results in a formula for the equilibrium film thickness on a superheated substrate

$$d = \left(\frac{T_{sat}|A|}{\rho \mathscr{L} \Delta T}\right)^{1/3}. \tag{6.13}$$

Since the Hamaker constant is negative, we replaced "$-A$" with the absolute value of A. According to (6.13), the equilibrium thickness increases as the superheat is decreased. Note, however, that taking the limit of $\Delta T \to 0$ is not physically meaningful since the van der Waals model of disjoining pressure is only applicable in a limited range of values of d and therefore cannot be used for very small ΔT. Equilibrium for $\Delta T = 0$ can be reached for any *macroscopic* value of the film thickness since for macroscopic films the disjoining pressure is negligible and therefore the chemical potentials of the two phases are independent of the film thickness.

Finally, let us discuss the regime in which the disjoining pressure is dominated by its electrostatic component. Using the Debye–Hückel approximation for the latter, derived in Sect. 4.3, and neglecting the van der Waals component, the equilibrium condition (6.12) is replaced with

$$2\varepsilon \lambda_D^{-2} \psi_0^2 e^{-2d/\lambda_D} = \frac{\rho \mathscr{L} \Delta T}{T_{\text{sat}}}, \tag{6.14}$$

where ε is the dielectric permittivity of the liquid, ψ_0 is the electric potential at the solid–liquid interface, and λ_D is the Debye length introduced in Chap. 4.

It is important to note that thermodynamic quantities describing the liquid film in general depend on the magnitude of the electric field $|\mathbf{E}|$ there. However, to establish local equilibrium condition at the interface, given by (6.14), the correction to the chemical potential due to disjoining pressure is evaluated near the liquid–vapor interface, i.e., in the region in which the electric field $\mathbf{E}$ can be assumed negligible. The latter conclusion is based on the boundary condition of zero normal component of $\mathbf{E}$ at the liquid–vapor interface and the fact that there is no electric field tangential to the interface.

The equilibrium film thickness can be expressed from (6.14) as:

$$d = \frac{\lambda_D}{2} \ln \left(\frac{2\varepsilon \psi_0^2 T_{\text{sat}}}{\rho \mathscr{L} \Delta T \lambda_D^2} \right). \tag{6.15}$$

Physically meaningful (i.e., positive) values of d are obtained only if the superheat is below a critical value determined by:

$$\Delta T^* = \frac{2\varepsilon \psi_0^2 T_{\text{sat}}}{\rho \mathscr{L} \lambda_D^2}. \tag{6.16}$$

For an aqueous solution at normal pressure with $\lambda_D = 10\,\text{nm}$, $\psi_0 = 0.1\,\text{V}$, and $\Delta T = 10^{-2}\,\text{K}$, the equilibrium thickness predicted by (6.15) is approximately 4.2 nm. At larger λ_D, higher values of d up to $\sim$50 nm are obtained for realistic superheats.

6.1.2 Equations for Evaporative Flux

Thermodynamic equilibrium at a vapor–liquid interface implies that there is no net mass flux between liquid and vapor, but it does not mean that there is no exchange of individual molecules at the interface. In fact, some molecules from the vapor strike the liquid–vapor interface and remain in the liquid phase afterward, while some molecules from the liquid constantly escape into the vapor phase. These two processes balance each other under the equilibrium conditions, which is the reason why a term "dynamic equilibrium" is often used to describe liquid–vapor systems at the saturation temperature. Let us define the equilibrium flux of molecules j_e as the number of molecules transferred from the vapor to the liquid per unit time per unit area of the interface (assumed flat). At equilibrium and for negligible disjoining pressure, both temperature T and pressure p are uniform throughout the liquid–vapor system and the pressure is equal to the equilibrium saturation value p_e corresponding to the temperature T. Under these conditions the flux j_e can be computed using the kinetic theory of gases as follows.

Consider a flat liquid–vapor interface which coincides with the y–z-plane so that the x-axis is normal to it and the vapor phase is an ideal gas occupying the region $x > 0$. Dimensional velocity components of individual vapor molecules are denoted by u, v, and w, and the velocity distribution function is denoted by $f(u,v,w)$, meaning that in a unit volume of vapor the average number of molecules with the x-velocity component between u and $u + du$, the y-velocity component between v and $v + dv$, and the z-velocity component between w and $w + dw$ is given by $f(u,v,w)dudvdw$. Even though quantities such as du are treated as differentials in the derivations below, the value of $f(u,v,w)dudvdw$ is assumed to be large enough so that the actual number of molecules with velocity components in the specified ranges at any given time is close to the average value. This is a standard approach in the kinetic theory of gases [84]. Knowing the velocity distribution in the vapor allows one to find the number of vapor molecules striking the liquid surface per unit time. Not all of these molecules end up in the liquid since some are reflected back into the vapor. The fraction of molecules striking the interface that in fact remain in the liquid phase afterward is the so-called accommodation coefficient $\hat{\sigma}$, determined empirically. The total flux of molecules transferred to the liquid is then given by:

$$j_e = - \int\limits_{-\infty}^{\infty} \int\limits_{-\infty}^{\infty} \int\limits_{-\infty}^{0} \hat{\sigma} u f(u,v,w)dudvdw. \tag{6.17}$$

Here, we accounted for the fact that for each value of u, only molecules at the distances less than udt can reach the interface within the time interval dt. We neglect the effect of collisions which may deflect molecules traveling toward the interface; this is justified regardless of the average collision frequency as long as one considers the limit of $dt \to 0$. The negative sign in front of the integral in (6.17) is needed since j_e is defined above as a positive quantity while integration is over negative values of

u (corresponding to molecules moving toward the vapor–liquid interface). Using the classical Maxwell velocity distribution for an ideal monatomic gas at rest, we write

$$f(u,v,w) = n \left(\frac{M}{2\pi RT} \right)^{3/2} e^{-\frac{M}{2RT}(u^2+v^2+w^2)},$$ (6.18)

where n is the total number density (the number of molecules per unit volume), M is the molar mass. Substituting the formula for $f(u,v,w)$ into (6.17) and integrating, we arrive at

$$j_e = \hat{\sigma} n \sqrt{\frac{RT}{2\pi M}}.$$ (6.19)

Here, we evaluated the integral over u using a standard substitution and the integrals over v and w using the formula

$$\int_{-\infty}^{\infty} e^{-\xi^2} d\xi = \sqrt{\pi}.$$ (6.20)

For an ideal gas, the value of n can be related to the Avogadro number N_A and the macroscopic parameters according to

$$\frac{n}{N_A} = \frac{p_e}{RT}$$ (6.21)

and therefore

$$j_e = \frac{\hat{\sigma} p_e N_A}{\sqrt{2\pi RTM}}.$$ (6.22)

Let us now suppose that the pressure in the system p is slightly below the value corresponding to the liquid–vapor coexistence, $p_e(T)$, while the temperature is equal to T. Under these conditions, liquid is expected to evaporate so that there is a net transfer of molecules from the liquid to the vapor. The flux of molecules emitted by the liquid, j^+, can be assumed equal to j_e despite the fact that the pressure in the liquid is not equal to the saturation value. Indeed, since the liquid is essentially incompressible, the change in pressure does not result in significant change of the number density of molecules on the liquid side, n_l. The value of n_l and the characteristic velocity of molecules (proportional to the square root of temperature) are the key parameters determining the flux j^+ regardless of the details of the velocity distribution function in the liquid, so one can assume $j^+ = j_e$.

The fact that the molecules emitted by the liquid have a Maxwellian velocity distribution characterizes only part of the overall vapor velocity distribution function, corresponding to $u > 0$, near the interface. The part of the distribution function corresponding to $u < 0$ can be rather complicated due to collisions between molecules coming from the bulk vapor and the ones escaping from the interface. However, there is a finite distance δ_K over which the interface effects are felt in the gas: if molecules escaping from the liquid reach the boundary of this region,

Fig. 6.2 Geometric
configuration used in the
derivation of the formulas for
the evaporative flux

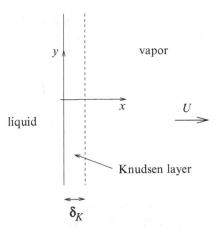

they already experienced many collisions with vapor molecules, so the information
about their original velocity distribution (coming out of the liquid) has been lost. It
is therefore natural to assume that for $x > \delta_K$ the velocity distribution is that of an
ideal gas moving at a bulk velocity of U away from the interface, i.e.,

$$f_b(u,v,w) = n_b \left(\frac{M}{2\pi RT} \right)^{3/2} e^{-\frac{M}{2RT}\left((u-U)^2 + v^2 + w^2 \right)}, \qquad (6.23)$$

where n_b is the number density in the bulk. Thus, the vapor phase is divided
into two regions as sketched in Fig. 6.2, the so-called Knudsen layer ($x < \delta_K$)
with a complicated velocity distribution, and the bulk region in which the velocity
distribution is given by (6.23). The thickness of the Knudsen layer is on the order
of the mean free path [78], typically below $1\,\mu$m. Note that the bulk velocity U is
not known in advance and has to be determined as part of the solution. In many
evaporation experiments, the nondimensional quantity

$$\phi = \frac{U}{\sqrt{2RT/M}} \qquad (6.24)$$

is small (since $\sqrt{RT/M}$ is of the order of $100\,$m/s) and therefore instead of f_b
defined by (6.23) one can use its linearized version,

$$\tilde{f}_b(u,v,w) = n_b \left(\frac{M}{2\pi RT} \right)^{3/2} \left(1 + \sqrt{\frac{2M}{RT}} \phi u \right) e^{-\frac{M}{2RT}\left(u^2 + v^2 + w^2 \right)}. \qquad (6.25)$$

The flux j_K^- of molecules entering the Knudsen layer from the right (at $x = \delta_K$) is found by integrating this function,

$$j_K^- = -\int\limits_{-\infty}^{\infty} \int\limits_{-\infty}^{\infty} \int\limits_{-\infty}^{0} u\tilde{f}_b(u,v,w)dudvdw, \tag{6.26}$$

resulting in the following simple formula

$$j_K^- = \frac{p_v N_A}{\sqrt{2\pi RTM}}\left(1 - \phi\sqrt{\pi}\right). \tag{6.27}$$

The integration procedure here is very similar to the one used in deriving (6.22) except that integration by parts is required to obtain the second term in the brackets on the right-hand side of (6.27). The ideal gas law was used to express n_b in terms of vapor pressure and temperature in the bulk. We now assume that the flux given by (6.27) multiplied by the accommodation coefficient $\hat{\sigma}$ is equal to the flux of molecules transferred from the vapor to the liquid, j^-, at the liquid–vapor interface ($x = 0$), as is done in most studies of evaporation. Then,

$$j^- = \hat{\sigma}\frac{p_v N_A}{\sqrt{2\pi RTM}}\left(1 - \phi\sqrt{\pi}\right). \tag{6.28}$$

Note, however, that it is not a priori obvious that the fluxes $\hat{\sigma}j_K^-$ and j^- should be equal to each other. In fact, their equality is an assumption best justified by the fact that more elaborate theories and numerical simulations, discussed below in Sect. 6.4*, indicate that the difference between their values is in fact rather small, at least in the limit of small ϕ.

The evaporative mass flux J at the interface is the difference between j^+ (flux of molecules from liquid to vapor $x = 0$, assumed equal to j_e) and j^- multiplied by the mass of a molecule, so by combining (6.22) and (6.28), we can write

$$J = \hat{\sigma}\frac{p_e - p_v(1 - \phi\sqrt{\pi})}{\sqrt{2\pi RT/M}}. \tag{6.29}$$

From the conservation of mass it is clear that $J = \rho U$, which together with (6.24) allows one to express the parameter ϕ in terms of J. Substituting this result into (6.29) together with the ideal gas law leads to an equation for the evaporative mass flux J at the liquid–vapor interface in the form

$$J = \frac{2\hat{\sigma}}{2 - \hat{\sigma}}\frac{p_e - p_v}{\sqrt{2\pi RT/M}}. \tag{6.30}$$

This formula is known as the Hertz–Knudsen–Schrage equation or the Kucherov–Rikenglaz equation. While derived for the case of evaporation, it is equally applicable to the case of condensation. For small $\hat{\sigma}$, it reduces to the classical Hertz–Knudsen equation (also sometimes called the Hertz–Knudsen–Langmuir or HKL equation),

$$J = \hat{\sigma}\frac{p_e - p_v}{\sqrt{2\pi RT/M}}. \qquad (6.31)$$

However, for values of $\hat{\sigma}$ near unity, using (6.31) instead of (6.30) leads to a significant error.

6.2 Steady Evaporating Meniscus

The notion of an apparent contact line for isothermal systems was introduced in Sect. 2.1.3. The term is justified by the fact that the interface appears to be in contact with a dry solid wall based on macroscale observations, but by gradually zooming into what looks like a contact line one discovers that there is actually a transition region between the macroscopically observed interface shape and an ultrathin film covering what appears to be a dry solid surface. Apparent contact lines are also often encountered in nonisothermal systems, except that the phase change processes occurring in such systems can have a significant effect on the actual shape of the interface in the transition region. We start by considering a steady configuration, i.e., assume that phase change processes and global fluid flow in the system are coupled together in such a way that the interface is stationary. Remarkably, it turns out that the *local* interface shape and evaporative flux near such stationary apparent contact lines for a given solid surface temperature are essentially the same for a wide range of geometric configurations encountered, e.g., in the studies of micro heat pipes, boiling in microchannels, and many other applications. This local solution structure was investigated in the pioneering works of Potash and Wayner [101] and Moosman and Homsy [92]. Here, we discuss this solution using the general framework employed in later studies [4, 93] and neglecting the effects of gravity.

Consider a two-dimensional steady configuration sketched in Fig. 6.3. The heated solid substrate is at a uniform dimensional temperature T_H^* above the vapor saturation temperature T_S^*. An ultrathin film shown in the sketch, often called an adsorbed film, covers the macroscopically dry part of the substrate. The film is stable due to the action of London–van der Waals forces, as discussed in detail

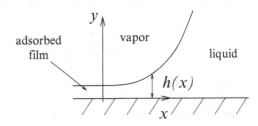

Fig. 6.3 Local geometric configuration in the transition region between a constant-curvature meniscus and an ultrathin adsorbed film

in Sect. 6.1.1, so its thickness is expressed in terms of the Hamaker constant A according to (6.13). The objective of the present section is to describe the transition region between this adsorbed film and the macroscopically observed meniscus. The latter is assumed to have a known constant-curvature, κ, far away from the solid substrate. In general, the interface shape in the transition region depends on fluid flow in both liquid and vapor. The flow equations in the two phases are coupled through several conditions at the liquid–vapor interface. The first one is conservation of mass across the interface,

$$\rho \mathbf{U}_l \cdot \mathbf{n} = \rho_v \mathbf{U}_v \cdot \mathbf{n}, \tag{6.32}$$

where ρ and ρ_v are the densities of the liquid and vapor, respectively, $\mathbf{U}_l$ is the liquid velocity at the interface, $\mathbf{U}_v$ is the vapor velocity evaluated just outside of the Knudsen layer (in which the macroscopic flow equations are expected to break down), and $\mathbf{n}$ is the normal vector to the liquid–vapor interface pointing into the vapor. The second interfacial condition is obtained from the standard normal stress balance for a two-fluid system without phase change, (1.104), by adding a term due to the difference in the momentum of the vapor flowing away from and the liquid flowing toward the liquid–vapor interface:

$$J^2 \left(\frac{1}{\rho_v} - \frac{1}{\rho} \right) + \mathbf{n} \cdot \left(\mathbf{T} - \mathbf{T}^{(v)} \right) \cdot \mathbf{n} = 2H\sigma + \Pi. \tag{6.33}$$

Here, J is the evaporative mass flux defined in the previous section, $\mathbf{T}$ and $\mathbf{T}^{(v)}$ are the stress tensors in the liquid and the vapor, respectively, H is the mean curvature, σ is the surface tension, Π is the disjoining pressure. The tangential stress balance (in the direction of a tangent vector $\mathbf{t}$) is

$$J(\mathbf{U}_l - \mathbf{U}_v) \cdot \mathbf{t} = \mathbf{t} \cdot \left(\mathbf{T} - \mathbf{T}^{(v)} \right) \cdot \mathbf{n} - \nabla \sigma \cdot \mathbf{t}. \tag{6.34}$$

While there can be a small jump in the tangential velocity across the Knudsen layer, we neglect it in the present analysis, resulting in the simplification of the tangential stress condition to the form

$$\mathbf{t} \cdot \left(\mathbf{T} - \mathbf{T}^{(v)} \right) \cdot \mathbf{n} = \nabla \sigma \cdot \mathbf{t}. \tag{6.35}$$

Finally, the energy balance at the liquid–vapor interface is written as:

$$J\mathscr{L} + \frac{J^3}{2} \left(\frac{1}{\rho_v^2} - \frac{1}{\rho^2} \right) = k_v \nabla T^{(v)} \cdot \mathbf{n} - k \nabla T^* \cdot \mathbf{n}, \tag{6.36}$$

with k (k_v) and T^* ($T^{(v)}$) denoting the thermal conductivities and temperatures in the liquid (vapor). Viscous dissipation at the interface can also contribute to the energy balance there [27], but this effect is neglected in the present formulation.

The density, thermal conductivity, and viscosity of the vapor are typically much smaller than those of the liquid, so it is natural to consider the limit when the corresponding ratios approach zero. However, since vapor velocity can be large, the product of ρ_v and the normal velocity of the vapor is assumed to be an order one quantity in our asymptotic limit. Taking the limit of $\frac{\rho_v}{\rho} \to 0$ in (6.33) and (6.36) implies that the terms proportional to ρ^{-1} and ρ^{-2} can be neglected. The assumption of small vapor viscosity (compared to that of the liquid) implies that the viscous part of the stress tensor in the vapor can be set to zero. Taking the limit of $\frac{k_v}{k} \to 0$ allows us to omit the first term on the right-hand side of (6.36). After all these simplifications, the interfacial conditions take the following form,

$$\frac{J^2}{\rho_v} + \mathbf{n} \cdot \mathbf{T} \cdot \mathbf{n} = 2H\sigma + \Pi - p_v, \qquad (6.37)$$

$$\mathbf{t} \cdot \mathbf{T} \cdot \mathbf{n} = \nabla \sigma \cdot \mathbf{t}, \qquad (6.38)$$

$$J\mathscr{L} + \frac{J^3}{2\rho_v^2} = -k\nabla T^* \cdot \mathbf{n}. \qquad (6.39)$$

The approximation defined by (6.32), (6.37), (6.38), and (6.39) is known as the *one-sided model* of evaporation and was formulated by Burelbach et al. [27]. While we consider only steady evaporation in the present section, the one-sided model in a slightly modified form is also applicable to moving interfaces.

The relative importance of vapor recoil, which is the first term in (6.37) and has to do with the acceleration of the flow in the interfacial region, for sufficiently high curvature is not significant compared to the interfacial capillary pressure jump. The latter can be estimated using the value of curvature away from the wall and is typically above $10\,\mathrm{N/m^2}$. The evaporative mass flux, e.g., in experiments with heptane on silicon substrate conducted by DasGupta et al. [41] $J \sim 10^{-3}\,\mathrm{kg/(m^2\,s)}$, which gives the estimate $J^2\rho_v^{-1} \sim 10^{-7}\,\mathrm{N/m^2}$, well below the capillary pressure jump at the interface. Similar estimates can be obtained for conditions corresponding to other well-known experimental studies of evaporating menisci, reviewed, e.g., by Wayner [136], thus justifying the simplification of (6.37) to the form

$$\mathbf{n} \cdot \mathbf{T} \cdot \mathbf{n} = 2H\sigma + \Pi - p_v. \qquad (6.40)$$

Since $\mathscr{L}$ is of the order of 10^5–$10^6\,\mathrm{J/kg}$ for most liquids, the second term on the left-hand side of (6.39) is very small compared to the first one in almost all evaporation experiments and therefore can be neglected, resulting in

$$J\mathscr{L} = -k\nabla T^* \cdot \mathbf{n}. \qquad (6.41)$$

Using (6.30) with $\hat{\sigma} = 1$, we obtain

$$\frac{J}{2}\sqrt{\frac{2\pi R T_s^*}{M}} = \rho_v \mathscr{L}\frac{T^i - T_s^*}{T_s^*} - \frac{\rho_v}{\rho}(2H\sigma + \Pi). \qquad (6.42)$$

Here we express p_e in terms of the local interfacial temperature T^i using the conditions of thermodynamic equilibrium together with (6.2).

In order to apply the one-sided model of evaporation to the geometric configuration shown in Fig. 6.3 it is natural to recast the model in nondimensional terms. Since the interface is steady, the flow in the liquid has to compensate for evaporative mass loss at the interface. Thus, the characteristic flow velocity is defined by the evaporation rate, which in turn can be estimated from the interfacial energy balance, (6.41). The natural velocity scale that arises from the energy balance is

$$V = \frac{kT_S^*}{\rho \mathscr{L} R_0}.$$ (6.43)

Here, R_0 is the characteristic length scale which is determined by the global geometry of the system and is of the order of the radius of curvature of the meniscus away from the solid. For example, when the global configuration is that of a bubble in a microchannel [4], R_0 is the cross-sectional dimension of the channel. The capillary number $Ca = \mu V / \sigma$ calculated based on (6.43) is typically very small.

Since the slope of the interface in the transition region shown in Fig. 6.3 is small (eventually approaching zero in the adsorbed film), it is natural to use different length scales in the directions along the solid (L_x) and normal to it (L_y), rather than scale all lengths by R_0. This allows us to utilize the lubrication-type approach which has already been proven useful in obtaining physically meaningful solutions in a number of cases discussed in the previous chapters. To determine L_x and L_y we follow the general procedure introduced in the derivation of the Landau–Levich scaling in Sect. 2.2, but with some modifications to account for the fact that the capillary number is based on V rather than the characteristic flow velocity in the direction along the solid. The latter, as well as the velocity scale in the vertical direction (V_{char}), can be expressed in terms of V, L_x, and L_y by making the following observations. Using (6.41) with the vertical length scale L_y, as appropriate for the transition region, we find

$$V_{char} = \frac{kT_S^*}{\rho \mathscr{L} L_y},$$ (6.44)

or, equivalently, $V_{char} = V R_0 / L_y$. The nondimensional velocity v in the vertical direction is then defined as the corresponding dimensional value divided by V_{char}. According to the standard argument in the development of the lubrication-type approximation, the ratio of velocity scales should be the same as that of the length scales. This can only be achieved if $L_x V_{char} / L_y$ is used as the scale for the nondimensional horizontal velocity component u. The scaled version of the momentum balance in the horizontal direction in the standard Stokes flow approximation for negligible gravity is

$$\frac{1}{R_0 L_x} p_x = Ca \left(\frac{R_0}{L_y^2 L_x} u_{xx} + \frac{L_x R_0}{L_y^4} u_{yy} \right).$$ (6.45)

Here, x and y are the Cartesian coordinates scaled by L_x and L_y, respectively, and shown in the sketch in Fig. 6.3, p is pressure scaled by σ / R_0. The derivation and conditions of applicability of the Stokes flow approximation are discussed in Sect. 1.5*. For small interfacial slope, $L_y \ll L_x$ and (6.45) can be simplified to

$$p_x = \mathrm{Ca} \frac{L_x^2 R_0^2}{L_y^4} u_{yy}. \tag{6.46}$$

Based on this form of the equation, a condition on the length scales is

$$\frac{L_y^2}{L_x R_0} \sim \mathrm{Ca}^{1/2} \quad \text{as} \quad \mathrm{Ca} \to 0. \tag{6.47}$$

Since the interface curvature has to be an order one quantity in the limit of small Ca to ensure matching to the constant value away from the solid, the length scales also have to satisfy

$$\frac{L_y}{L_x^2} \sim \frac{1}{R_0} \quad \text{as} \quad \mathrm{Ca} \to 0. \tag{6.48}$$

The length scales which satisfy both (6.47) and (6.48) are

$$L_x = \mathrm{Ca}^{1/6} R_0, \qquad L_y = \mathrm{Ca}^{1/3} R_0. \tag{6.49}$$

With these scales, (6.46) becomes

$$p_x = u_{yy} \tag{6.50}$$

and the interfacial stress conditions, (6.38) and (6.40) at leading order in the capillary number Ca are reduced to the following nondimensional forms,

$$u_y = 0, \tag{6.51}$$

$$p - \hat{p}_v = -h_{xx} - \frac{\varepsilon}{h^3}. \tag{6.52}$$

Here, we represent the interface shape by the function $y = h(x)$ and neglect the effects of thermocapillarity; $\hat{p}_v = p_v R_0 / \sigma$. The disjoining pressure is dominated by the London–van der Waals dispersion forces and is therefore characterized by the Hamaker constant A, as discussed in Sect. 1.7. The parameter ε is the nondimensional version of the Hamaker constant and is defined by:

$$\varepsilon = \frac{|A|}{\sigma \, \mathrm{Ca} \, R_0^2}. \tag{6.53}$$

The dimensional condition of conservation of mass, (6.32), suggests that the evaporative mass flux is of the order of ρV_{char}, leading to the definition of the corresponding nondimensional quantity

$$\hat{J} = \frac{J}{\rho V \mathrm{Ca}^{-1/3}}. \tag{6.54}$$

The condition of the interfacial energy balance is then

$$\hat{J} = -T_y, \tag{6.55}$$

where T is the temperature scaled by T_S^*. The nondimensional version of the condition of conservation of mass, (6.32), is

$$\hat{J} + uh_x - v = 0. \tag{6.56}$$

Finally, the equation for the scaled mass flux obtained using (6.42) and (6.55), is in the form

$$\hat{J} = \frac{T_H - 1 - \delta(h_{xx} + \varepsilon h^{-3})}{K + h}, \tag{6.57}$$

where $T_H = T_H^*/T_S^*$; K and δ are given by:

$$K = \frac{T_S^* Ca^{-1/3}}{2\rho_v \mathscr{L}^2 R_0} \sqrt{2\pi R T_S^*/M}, \qquad \delta = \frac{\sigma}{\mathscr{L}\rho R_0}. \tag{6.58}$$

The velocity profile $u = u(y)$ is obtained by integrating (6.50) with the no-slip condition at the solid and the zero shear-stress condition at the interface, resulting in

$$u = \frac{P_x}{2}(y^2 - 2yh). \tag{6.59}$$

We then follow the usual steps of the development of the lubrication-type models and substitute the velocity profile into the integral mass balance, resulting in the following equation for the interface shape,

$$\frac{\delta h_{xx} + \delta \varepsilon h^{-3} - T_H + 1}{K + h} = \frac{1}{3} \left[h^3 \left(h_{xx} + \varepsilon h^{-3} \right)_x \right]_x. \tag{6.60}$$

This equation has to be solved with the condition of matching to the known curvature of the meniscus away from the wall,

$$h''(\infty) = R_0 \kappa, \tag{6.61}$$

and the condition of *all* derivatives approaching zero in the flat adsorbed film region of specified thickness h_{af}. The latter is found from (6.57) based on the condition of no evaporation ($\hat{J} = 0$) in the flat film, resulting in the expression

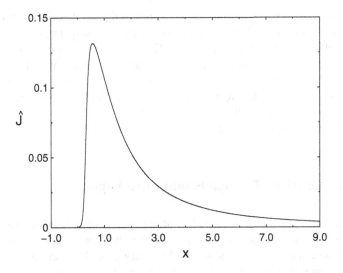

Fig. 6.4 Scaled evaporative flux as a function of the local coordinate for $T_H = 1.4$, $K = 2$, $\delta = 0.04$, $\varepsilon = 10^{-3}$. From [4], reprinted with permission of Elsevier

$$h_{af} = \left(\frac{T_H - 1}{\delta\varepsilon}\right)^{-1/3}. \tag{6.62}$$

Following the same general approach as, e.g., in Sect. 3.3, we consider the equation for the interface shape on a finite domain $[0, L]$ such that near $x = 0$

$$\zeta \equiv \frac{h - h_{af}}{h_{af}} \ll 1 \tag{6.63}$$

and thus the boundary conditions at $x = 0$ can be formulated based on the linearized equation for ζ. Two constants appear in the boundary conditions, one of which can be chosen as an arbitrary small number, while the second one is determined from the matching condition. The numerical solution based on MATLAB bvp4c subroutine is described in [103]. MATLAB code for this problem can be found in Sect. B.6 and allows one to obtain numerical solutions for the interface shape over a range of parameters. For practical applications, an important quantity is the scaled evaporative flux $\hat{J}$, which can be found using (6.57) once the numerical interface shape is obtained. The result is shown in Fig. 6.4. The flux clearly has a maximum in the transition region, meaning that most of the evaporation takes place near the apparent contact line. This observation is useful for practical applications since by increasing the length of the apparent contact line one can increase the efficiency of heat transfer, which is a typical goal in applications.

It is interesting to note that local solutions such as the one described in the present section can be incorporated into complicated geometric configurations, e.g., a three-dimensional bubble in a microchannel [4]. The local solutions in this configuration have to be matched asymptotically to the capillary static shapes which are assumed to be independent from phase change processes and were discussed in detail in Chap. 3. Application of this approach results in the prediction of a nearly linear growth of vapor bubble length as a function of heating intensity, which was indeed observed in experiments of Yang and Homsy [145].

6.3 Evaporation of Droplets into Pure Vapor

The lubrication-type approach to studies of apparent contact lines discussed in the previous section can be generalized to describe situations when the interface is no longer stationary, as was first shown by Ajaev et al. [6]. Several later studies applied this approach to moving contact line problems. Here, we consider a model problem of a droplet spreading and evaporating on a heated solid substrate using the general mathematical framework for studies of evaporating droplets developed by Anderson and Davis [9]. Since our objective is to illustrate the approach, we consider two simplifying assumptions: constant uniform temperature of the substrate and negligible thermocapillary effects. A more elaborate model which incorporates the effects of thermocapillarity and heat conduction in the solid is developed and compared with experiments in Sodtke et al. [117].

The problem geometry is illustrated in Fig. 6.5. Since the droplet is axisymmetric, we use cylindrical coordinates r and z, shown in the sketch. The radial variable is scaled by the characteristic wetted radius R_0 and the vertical one by $Ca^{1/3}R_0$; the capillary number Ca is based on the velocity V defined in the previous section. The length scales are found by using the same arguments as in the previous section and result in the simplified lubrication-type system of equations of the form

$$(ru)_r/r + w_z = 0, \tag{6.64}$$

$$-p_r + u_{zz} = 0, \tag{6.65}$$

$$p_z = 0, \tag{6.66}$$

$$T_{zz} = 0. \tag{6.67}$$

The scaled velocity components in the radial direction and the vertical direction are denoted by u and w, respectively. The nondimensional temperature T is defined in terms of its dimensional value, T^*, according to

$$T = \frac{T^* - T_S^*}{Ca^{2/3}T_S^*}. \tag{6.68}$$

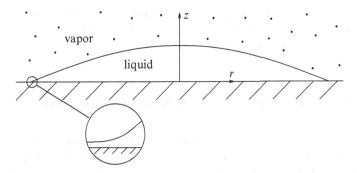

Fig. 6.5 Sketch of an evaporating droplet on a uniformly heated substrate with an enlarged view of the transition between the droplet and the adsorbed film

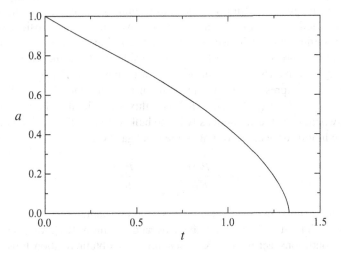

Fig. 6.6 Wetted radius of an evaporating axisymmetric droplet on a uniformly heated surface

The interfacial boundary conditions are those of the simplified one-sided model of evaporation except that time-dependent terms have to be added. Following the usual steps in the derivation of the lubrication-type model we obtain an equation for time-dependent droplet thickness $h(r, t)$ in the form

$$h_t - \frac{\delta \left[h_{rr} + r^{-1}h_r + \varepsilon h^{-3} \right] - T_0}{K + h} + (3r)^{-1} \left[rh^3 \left(h_{rr} + r^{-1}h_r + \varepsilon h^{-3} \right)_r \right]_r = 0.$$

(6.69)

This equation with two symmetry conditions at $r = 0$, fixed value of $h = h_{af}$ in the adsorbed film and zero first derivative there is solved numerically using the DVODE package. The wetted radius of the droplet, defined by the location of the maximum of the interfacial curvature, is plotted as a function of time in Fig. 6.6. The wetted radius initially decays in a nearly linear fashion, but as the droplet becomes smaller, the evaporation rate accelerates significantly. This can be explained by the observation that most of evaporation in this geometric configuration takes place in the apparent contact line region.

6.4* More Accurate Approximations for Evaporative Flux

The derivation of (6.30) for the evaporative flux at the liquid–vapor interface presented in Sect. 6.1.2 was based on a number of simplifying assumptions. The objective of the present section is to review studies which relax some of these assumptions and thus obtain expressions for the evaporative flux that have a wider range of applicability. First, we note that since during evaporation the liquid–vapor system is no longer in thermodynamic equilibrium, the temperature and pressure are not expected to be uniform throughout the system. Let us consider the same geometric configuration as in Sect. 6.1.2, sketched in Fig. 6.2, and denote the vapor temperature and pressure in the bulk (just outside of the Knudsen layer) by T_b and P_b, respectively. The liquid surface is at a temperature T_s and pressure P_s. It is convenient to express the results in terms of the nondimensional bulk velocity $\phi = \beta_b U$, where $\beta_b = \sqrt{M/2RT_b}$. The mass flux J can be then be obtained from $J = \rho_b U$, where ρ_b is the vapor density in the bulk, different from the vapor density ρ_s near the liquid surface. According to the ideal gas law,

$$\rho_b = \frac{P_b M}{RT_b}, \qquad \rho_s = \frac{P_s M}{RT_s}, \tag{6.70}$$

where M is the molar mass, R is the universal gas constant.

Following Ytrehus [146] and Labuntsov and Kryukov [81], we use the conservation conditions across the Knudsen layer to obtain relationships between temperatures and pressures in the bulk and at the surface. The bulk velocity distribution is described by the function f_b introduced in Sect. 6.1.2 and written in the notation of the present section as:

$$f_b(u,v,w) = n_b \pi^{-3/2} \beta_b^3 e^{-\beta_b^2 \left((u-U)^2 + v^2 + w^2\right)}. \tag{6.71}$$

The flux of molecules entering the Knudsen layer from the right, j_K^-, can be calculated using the same approach as in Sect. 6.1.2 except that the scaled velocity $\phi = \beta_b U$ is no longer assumed small, so

$$j_K^- = -\int\limits_{-\infty}^{\infty} \int\limits_{-\infty}^{\infty} \int\limits_{-\infty}^{0} u f_b(u,v,w)\,du\,dv\,dw = -\frac{n_b \beta_b}{\sqrt{\pi}} \int\limits_{-\infty}^{0} u e^{-\beta_b^2 (u-U)^2}\,du. \tag{6.72}$$

Introducing a new integration variable $\tilde{u} = \beta_b(u - U)$, we observe that

$$\int u e^{-\beta_b^2(u-U)^2} \, du = \frac{1}{2\beta_b^2} \int e^{-\tilde{u}^2} \, d\tilde{u}^2 + \frac{U}{\beta_b} \int e^{-\tilde{u}^2} \, d\tilde{u}, \qquad (6.73)$$

resulting in the following formula for the flux j_K^-,

$$j_K^- = \frac{n_b}{2\sqrt{\pi}\beta_b} F, \qquad F = e^{-\phi^2} - \phi\sqrt{\pi}(1 - \operatorname{erf} \phi). \qquad (6.74)$$

In Sect. 6.1.2, it was assumed that the flux j_K^- (multiplied by the accommodation coefficient $\hat{\sigma}$) is the same as the flux of molecules entering the liquid at $x = 0$. However, it was already pointed out that this assumption is difficult to justify since the velocity distribution f_0 at $x = 0$ is different from f_b. The simplest approach to account for the difference is to use the distribution function

$$f_0 = \begin{cases} f_s, & u \geq 0 \\ Bf_b, & u < 0 \end{cases}, \qquad (6.75)$$

where B is a constant to be determined and

$$f_s = n_s \pi^{-3/2} \beta_s^3 e^{-\beta_s^2(u^2+v^2+w^2)}. \qquad (6.76)$$

Here $\beta_s = \sqrt{M/2RT_s}$. Using the function f_0 instead of the bulk distribution function implies that the flux j^- is equal $B\hat{\sigma} j_K^-$, while the flux j^+ is still found by integrating f_s, as outlined in Sect. 6.1.2. The mass conservation condition then leads to

$$\frac{1}{2\sqrt{\pi}} \left(\frac{n_s}{\beta_s} - \frac{n_b BF}{\beta_b} \right) = n_b U. \qquad (6.77)$$

For simplicity, we assume $\hat{\sigma} = 1$ here and below. Using the definitions of β_b, β_s, and (6.70), (6.77) can be written as:

$$\frac{P_s}{P_b} \left(\frac{T_b}{T_s} \right)^{1/2} = BF + 2\sqrt{\pi}\phi. \qquad (6.78)$$

The momentum and energy flux for molecules entering the Knudsen layer from the right can be found by integrating the corresponding quantities per molecule multiplied by f_b. The details of the calculation are rather tedious, but once all integrals are evaluated, the conditions of conservation of momentum and energy lead to two equations for the unknown quantities T_b/T_s, P_b/P_s, B, and ϕ. Together with (6.78), these form a system of three equations for four unknowns, meaning that we can now obtain relationships between ϕ and the ratios T_b/T_s or P_b/P_s. The resulting formulas are

$$\left(\frac{T_b}{T_s}\right)^{1/2} = -\frac{\sqrt{\pi}}{8}\phi + \left(1 + \frac{\pi}{64}\phi^2\right)^{1/2}, \tag{6.79}$$

$$\frac{P_s}{P_b} = 2e^{-\phi^2}\left[F + G\left(\frac{T_b}{T_s}\right)^{1/2}\right]^{-1}, \tag{6.80}$$

where F is defined above in (6.74) and G is given by:

$$G = (1 + 2\phi^2)\left[1 - \mathrm{erf}\,\phi\right] - 2\pi^{-1/2}\phi e^{-\phi^2}. \tag{6.81}$$

In the work of Rose [108], the distribution function is chosen as:

$$f_0 = \begin{cases} f_s, & u \geq 0 \\ (1 + Bu)f_b, & u < 0 \end{cases}, \tag{6.82}$$

where u is the x-component of velocity. The calculation with this assumption results in rather complicated formulas, listed in [108] and not reproduced here.

An alternative approach is to solve the equation for the distribution function equation, as discussed, e.g., in Sone [118]. Typically, the so-called Boltzmann–Krook–Welander equation (BKW numerically, also called BGK equation in some studies, after Bhatnagar, Gross, and Krook) is used, which is essentially the classical Boltzmann equation with a specific simple form of the term describing collisions between molecules. The numerical results turn out to be in good agreement with the models of Ytrehus [146], Labuntsov and Kryukov [81] and in excellent agreement with the model of Rose [108], as seen in Fig. 6.7. The long-dashed line shows results based on the model of Schrage [113] which assumes the distribution function at the surface to be independent from the bulk parameters, in the form

$$f_0 = \begin{cases} f_s, & u \geq 0 \\ (1 + Bu)f_s, & u < 0 \end{cases}. \tag{6.83}$$

Even though the model of Schrage [113] has been widely used in heat transfer literature, Fig. 6.7 suggests that its predictions significantly deviate from the numerical results.

Finally, we note that the evaporation and condensation fluxes here are assumed independent, which is not always the case in experiments. The corrections due to the coupling between the two are discussed in [135].

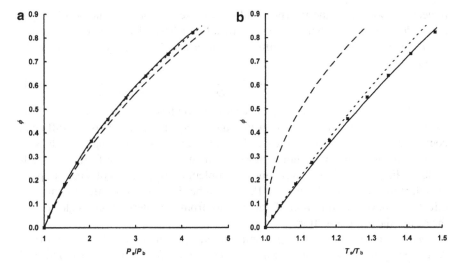

Fig. 6.7 Comparison between predictions of evaporation models of Rose [108] (*solid lines*), Ytrehus–Labuntsov–Kryukov [81, 146] (*dashed lines*), Schrage [113] (*long-dashed lines*), and numerical results based on the methods described in [118] (*filled squares*). From [108], reprinted with permission from Elsevier

6.5 Evaporation into Air

6.5.1 Modeling of Vapor Diffusion

In many applications of two-phase flows, one of the phases is a pure liquid and the other one is a mixture of air and vapor of the same liquid. For negligible disjoining pressure, thermodynamic equilibrium between such two phases is reached when both temperature and pressure are uniform and the partial pressure of the vapor is equal to its saturation value, p_{sat}. The latter can be expressed as a function of temperature using empirical formulas. For water vapor, the saturation pressure is equal to 2.31 kPa at 20°C and several empirical equations for $p_{sat}(T)$ are found, e.g., in Buck [26].

Consider a macroscopic uniform film of water initially in contact with isothermal air containing water vapor at uniform partial pressure p_v below the saturation pressure p_{sat}. The ratio p_v/p_{sat} is referred to as the relative humidity, denoted by H. While we are discussing evaporation of water in the present section, the results can be applied to evaporation of other pure liquids, e.g., ethanol or acetone. However, the value of H is then typically set to zero since initially air is not expected to contain any vapors of these liquids. The gas phase near the interface rapidly becomes saturated with the vapor, reaching the partial pressure of p_{sat}. After that, the rate of evaporation is determined by how fast the vapor is transported away from the interface. The two principal mechanisms for this process are diffusion of vapor molecules away from the interface and convective transport (meaning that vapor molecules are carried from the interface by a flow). The relative importance of the latter mechanism compared to diffusion is determined by the value of the Peclet number,

$$Pe = \frac{L_{char}V_g}{D},\qquad (6.84)$$

where L_{char} is the length scale, V_g the characteristic flow velocity in the gas phase (which can be estimated, e.g., from Bernoulli's equation), and D is the diffusion coefficient for vapor in air, see e.g., [102] for details. A wide range of values of Pe is encountered in applications. However, in situations relevant for many small-scale applications, e.g., in a well-studied problem of evaporation of small droplets on solid substrates [11, 43, 46], the Peclet number is small. Motivated by this, the effects of convective transport of vapor away from the interface are neglected for the rest of the present section.

Evaporation of a macroscopic uniform liquid film of initial thickness d is described by its thickness $h(t)$ (scaled by d). Since diffusion is the main physical effect determining the evaporation rate, the natural scale for the time variable t is given by L_{char}^2/D, where L_{char} is the characteristic length scale of change of the vapor concentration. The latter is found by solving the diffusion equation,

$$c_t = c_{zz}.\qquad (6.85)$$

Here, the respective scales for the coordinate z and the concentration c are L_{char} and c_{sat} (the saturation concentration). The coordinate z is in the direction normal to the solid surface and thus to the liquid–gas interface. It is convenient to describe vapor transport using concentration rather than partial pressure. The two quantities are proportional to each other, so the value of c far away from the interface is determined by the relative humidity,

$$c \to H \qquad \text{as} \qquad z \to \infty.\qquad (6.86)$$

We note that while the time derivative on the left-hand side of (6.85) is often negligible [43, 46], we keep it here because the initial stage of evaporation is expected to be rapid due to significant concentration gradients; the same approach is taken in, e.g., [11]. The condition of low Peclet number does not in general guarantee that the time-derivative of concentration in the diffusion equation is small.

The film thickness h (scaled by d) is coupled to the nondimensional concentration through the condition of conservation of mass at the interface stating that the evaporative mass loss from the liquid film per unit area per unit time $(\rho dD L_{char}^{-2} h_t)$ balances the vapor diffusion flux away from the interface $(D c_{sat} L_{char}^{-1} c_z)$. This condition suggests a natural definition of L_{char} as $\rho d/c_{sat}$, resulting in the simple nondimensional condition

$$h_t = c_z.\qquad (6.87)$$

If the dimensional film thickness is very small compared to the value of L_{char}, we can apply this boundary condition, as well as the condition of $c = 1$, at $z = 0$ instead of the time-dependent interface location, $z = h(t)$. To complete the mathematical formulation, we specify the initial conditions (assuming $z > 0$ here) in the form

$$h(0) = 1, \quad c(0,0) = 1, \quad c(z,0) = H. \tag{6.88}$$

Since the equation for c is formulated on a semi-infinite spatial domain and the boundary condition is that of constant concentration, we look for a similarity solution in the form

$$c = f(\eta), \quad \eta = \frac{z}{\sqrt{t}}. \tag{6.89}$$

By substituting this similarity ansatz into (6.85), we immediately obtain

$$f'' + \frac{1}{2}\eta f' = 0. \tag{6.90}$$

Integrating this equation twice with the conditions $f(0) = 1$ and $f(\infty) = H$, we arrive at the time-dependent concentration profile expressed in terms of the error function,

$$c = 1 + (H - 1)\mathrm{erf}\frac{z}{2\sqrt{t}}. \tag{6.91}$$

The film thickness as a function of time is then found from (6.87) with c_z evaluated from the similarity solution (6.91) and the initial condition given by $h(0) = 1$, resulting in the formula

$$h = 1 + \frac{2}{\sqrt{\pi}}(H - 1)\sqrt{t}. \tag{6.92}$$

Setting h to zero gives an estimate of the total scaled evaporation time,

$$t_e = \frac{\pi}{4(1 - H)^2}, \tag{6.93}$$

plotted as a function of humidity in Fig. 6.8. The fastest evaporation rate is achieved for zero relative humidity, i.e., the case of dry air. The evaporation time becomes very large when the humidity is close to $H = 1$ since the concentration gradient in this regime is small.

The dimensional values corresponding to t_e even for the case of dry air are very large, indicating that diffusion-limited evaporation is a very slow process. Shorter evaporation times are expected and in fact observed in large scale systems, i.e., during evaporation of a liquid layer on a kitchen counter, due to the effects of convective motion of air.

The physical mechanisms of evaporation of thin liquid films into air have been investigated in the works of Colinet and Haut [35] and Sultan et al. [126]. Models of vapor diffusion have been successfully applied to a number of geometric configurations, most notably the problem of evaporation of a small axisymmetric liquid droplet on a solid substrate [11, 43]. The results are shown to be in good agreement with the experimental data.

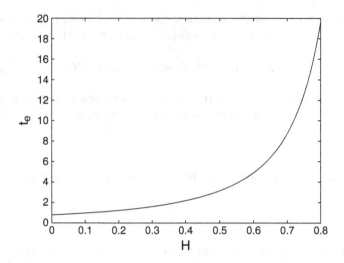

Fig. 6.8 Nondimensional time of diffusion-controlled evaporation of a uniform liquid film versus relative humidity

6.5.2   UltraThin Films in Contact with Air

In the previous section, we considered macroscopic films on solid substrates. For sufficiently small film thickness d, both the equilibrium conditions and film dynamics will be affected by disjoining pressure. Using the same approximation for disjoining pressure as in Subsect. 6.1.1, we can write

$$\mu_1 + \frac{A v_1}{d^3} = \mu_v. \tag{6.94}$$

Since the gas phase can be treated as a mixture of ideal gases, the chemical potential of the vapor in terms of its partial pressure p_v is

$$\mu_v = \mu_v^0 + RT \ln p_v, \tag{6.95}$$

where μ_v^0 is a function of temperature T only [79]. Based on the equilibrium conditions at negligible disjoining pressure, the value of μ_1 is equal to the chemical potential of the vapor at the saturation partial pressure, so μ_1 can be expressed as:

$$\mu_1 = \mu_v^0 + RT \ln p_{\text{sat}}. \tag{6.96}$$

Substituting the formulas for μ_v and μ_1 into (6.94) results in

$$\frac{A v_1}{d^3} = RT \ln \frac{p_v}{p_{\text{sat}}}. \tag{6.97}$$

Based on this equation, the equilibrium film thickness can be expressed in terms of the relative humidity $H = p_v/p_{sat}$ as:

$$d_{eg} = \left(\frac{Av_1}{RT \ln H} \right)^{1/3}. \tag{6.98}$$

Typical numerical values of d_{eg} from this formula are much smaller than the estimates for the pure vapor case based on (6.13) for comparable pressures and temperatures, indicating that such films are more difficult to detect experimentally.

6.5.3 Coupling Between Diffusion and Heat Transfer

In many practical applications, the solid substrate is heated to a temperature which is higher than the ambient moist air temperature T_a away from the interface. The dynamics of the film is then no longer described by the simple form given by (6.92) and depends on the coupled effects of diffusion and heat transfer. In this subsection, we follow [3] and investigate this coupling in a framework which also incorporates the effects of disjoining pressure. Since our objective is only to illustrate the role of different physical effects, the discussion will be limited to evaporation of a layer of uniform thickness. Even for this simple geometry, no analytical solutions are available, so the dynamics of the liquid film will be studied numerically.

The equation for vapor concentration, (6.85), is considered on a large but finite domain $[0, L]$, with the condition of fixed humidity H away from the interface applied at $z = L$. The definitions of nondimensional length, time, and concentration are the same as in Sect. 6.5.1. The condition for concentration at the air–liquid interface has to be modified to account for the dependence of the local saturation concentration on both disjoining pressure and local temperature. Using a linear dependence of the saturation concentration on temperature (following, e.g., Dunn et al. [46]) and the model of disjoining pressure from Subsect. 6.1.1, we obtain

$$c(0,t) = [1 + \alpha(T_{la} - 1)] \exp\left(-\frac{\varepsilon}{T_{la}h^3} \right), \tag{6.99}$$

where $\varepsilon = v_1|A|/(RT_a d^3)$, A is the Hamaker constant, T_{la} is the nondimensional liquid–air interface temperature. All temperature variables in this subsection are scaled by T_a.

The scaled temperature in the liquid is found by solving the quasi-steady equation

$$T_{\hat{z}\hat{z}}^1 = 0, \tag{6.100}$$

on the domain $[0, h]$, where $\hat{z}$ is the vertical coordinate scaled by d. We start with an assumption of fixed solid temperature, T_{sl}, thus providing a simple boundary condition for T^1. Heat transfer in the liquid is coupled to vapor diffusion through the condition of conservation of energy at the interface, written in the form

$$E T_{\hat{z}}^1 = c_z, \qquad E = \frac{k_1 T_a \rho}{\mathscr{L} D c_{sat}^2}, \tag{6.101}$$

where k_1 is the thermal conductivity of the liquid. The values of the evaporation parameter E are typically large ($E \sim 10^6$). To complete the problem formulation, we note that the film thickness $h(t)$ is still determined by (6.87) and the initial conditions are given by (6.88).

The solution of the quasi-steady equation (6.100) is written as:

$$T^1 = T_{sl} + \frac{T_{la} - T_{sl}}{h} \hat{z}, \tag{6.102}$$

where T_{la} is not known and has to be determined based on the numerical solution of the equation for concentration, which is found using a finite-difference approach. The vector of unknowns is formed by combining the concentration values at equally spaced mesh points on the interval $[0, L]$ and the additional two unknowns, T_{la} and h. Equations (6.101) and (6.87) are added to the standard equations for concentration values at mesh points based on (6.85). The DVODE solver is used to advance the solution in time.

Typical simulation results for the film thickness are presented in Fig. 6.9. The solution corresponding to $T_{sl} = 1$ is shown by the dashed line for comparison. We observe that both solutions approach constant values for large times, corresponding to conditions when the evaporation is suppressed by disjoining pressure. The equilibrium film thickness is lower for higher substrate temperature, according to the formula

$$h_{eq} = \left(\frac{\varepsilon}{T_{sl}}\right)^{1/3} \left[\ln\left(\frac{1 + \alpha(T_{sl} - 1)}{H}\right)\right]^{-1/3}, \tag{6.103}$$

which is obtained by setting all time-derivatives in our model to zero.

By comparing the two curves in Fig. 6.9, it is clear that the initial evaporation rate is larger when the substrate is heated, since in this case the concentration gradient is higher than for the nearly isothermal situation. This is due to the fact that the local equilibrium vapor concentration near the interface increases with temperature.

In experiments, the temperature of the solid–liquid interface is often not fixed and in fact depends on heat conduction in the solid substrate. For example, in the experiments of Sodtke et al. [117] the solid consists of two main layers: an electrically heated steel foil and a layer of Plexiglas. The thickness of the latter is much larger than that of the foil, so for heat conduction there we use the same

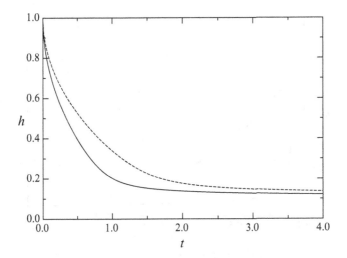

Fig. 6.9 Solutions for thickness using van der Waals disjoining pressure ($\varepsilon = 10^{-3}$) with fixed solid substrate temperature of $T_{\mathrm{sl}} = 1.01$ (*solid line*) and with the effects of substrate heating neglected (*dashed line*). From [3], reprinted with permission

length scale as for the processes in the vapor, i.e., we write the equation in terms of the nondimensional variable z. The small thickness of the foil motivates the choice of the length scale there as equal to d and the neglect of the unsteady term in the heat conduction equation. The equations then take the form

$$T_{\hat{z}\hat{z}}^{\mathrm{f}} = -q, \qquad -h_{\mathrm{f}} \leq \hat{z} \leq 0, \tag{6.104}$$

$$T_t^{\mathrm{P}} = \kappa T_{zz}^{\mathrm{P}}, \qquad -w \leq z \leq 0. \tag{6.105}$$

Here, T^{f} and T^{P} denote the nondimensional temperature profiles in the foil and the Plexiglas layer, respectively, k is the scaled thermal diffusivity of plexiglas and q is the heat source density in the foil used to describe the electric heat generation there. The scaled thicknesses of the foil and the Plexiglas layer are denoted by h_{f} and w, respectively. The equations for temperature are solved with the conditions of continuity of temperature and heat flux at the liquid–foil and foil–Plexiglas interfaces.

The numerical solution of the coupled system for heat conduction and vapor diffusion is similar to the one described earlier in the present subsection except that a finite-difference solution of the unsteady heat equation in the solid is added. Typical results for $w = 0.04$, $h_{\mathrm{f}} = 1$, and $\varepsilon = 10^{-3}$ are shown in Fig. 6.10. The evaporative flux is significantly higher than predicted by the fixed-substrate temperature model. In both cases, the evaporation rates are rapidly reduced as the effects of disjoining pressure become significant.

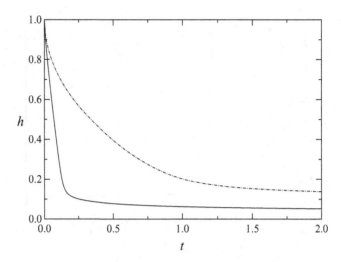

Fig. 6.10 Thickness of the liquid film as a function of time in nondimensional variables found from the model with heat transfer in the substrate (*solid line*) and the model with fixed substrate temperature (*dot–dashed line*). From [3], reprinted with permission

6.6   Notes on Literature

A range of topics related to phase change phenomena is covered in the book of Carey [30]. The methods of rarified gas dynamics which can be used to better understand the dynamics of vapor phase are discussed in several classical books, e.g., Kogan [78], and Sone [118]. Both theoretical and experimental studies of steady evaporating menisci are reviewed by Wayner [136]. Extensions of this approach to moving contact lines and its applications to bubbles in microchannels are discussed in Ajaev and Homsy [5].

Chapter 7
Flows with Surfactants

7.1 Surfactant Spreading on Liquid Film

Fluid flow on microscale can be induced and controlled efficiently by controlling the properties of the interfaces, as illustrated by a number of examples in the previous chapters, e.g., in the discussions of thermocapillary and electroosmotic flows. It is natural to ask if the properties of liquid–gas interfaces, most notably the surface tension, can be controlled by changing the chemical composition of the liquid, e.g., by adding a small amount of a chemical impurity (molecules different from those of the liquid) to an otherwise pure liquid. If the fraction of chemical impurity molecules is small and their equilibrium spatial distribution in the liquid is approximately uniform, they are not expected to have a significant effect on surface tension, since the overwhelming majority of the molecules in the interfacial region are still those of the liquid and not of the impurity. However, it turns out that the seemingly natural assumption of uniform spatial distribution of the impurity molecules is not applicable for a large class of chemicals known as surface active agents, or *surfactants*. For these chemicals, the equilibrium configuration is when nearly all the molecules in the interfacial region are actually at the interface itself, resulting in a dramatic change of the value of the surface tension compared to that of pure liquid. While we focus on gas–liquid (mostly air–water) interfaces in this chapter, similar phenomena can be observed at liquid–liquid interfaces when one of the liquids is polar, meaning that its molecules have nonzero electric dipole moment, and the other one is not, e.g., at water–oil interfaces.

A typical surfactant molecule added to water is elongated and consists of a large number of atoms with a hydrophobic (water-repellant) chemical group at one end, called the hydrophobic tail, and a hydrophilic group at the other end, called the hydrophilic head. This chemical structure suggests that it would be energetically favorable for such molecule to position itself at an air–water interface with the "head" inside the water and the "tail" sticking out into the air phase. Thus, surfactant molecules tend to accumulate at the air–water interface, forming a monolayer. Motivated by experiments in which nearly all of surfactant molecules are at the

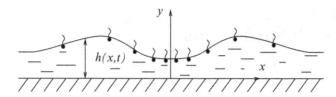

Fig. 7.1 Liquid–air interface deforming under the action of surfactants; nondimensional Cartesian coordinates are also shown

interface rather than in the bulk, we consider the limiting case of the so-called insoluble surfactant, i.e. assume that the surfactant concentration in the bulk is identically zero everywhere. The evolution of surfactant distribution is described by an equation for its interfacial concentration Γ^*, measured in units of mass divided by area. The surface tension σ under the isothermal conditions is a function of surfactant concentration and for sufficiently small Γ^* can always be approximated by a linear function,

$$\sigma = \sigma_0 - \gamma_\Gamma \Gamma^*, \tag{7.1}$$

where σ_0 is the surface tension of pure water, and γ_Γ is a constant which depends on the properties of the surfactant. The linearized equation (7.1) turns out to be sufficiently accurate for values of Γ^* encountered in a number of practical applications involving air–water interfaces and can also be applied to other gas–liquid and liquid–liquid interfaces if σ_0 is interpreted as the surface tension at the same interface without surfactants; the value of γ_Γ then depends on the properties of the surfactant and the liquid(s) involved.

Variations in the surfactant concentration generate shear stresses and thus induce liquid flow near the interface, directed from the regions of higher concentration to the regions of lower concentration when γ_Γ is positive (which is the usual case). The characteristic velocity of this flow, U (specified below), is used in the definition of the capillary number,

$$Ca = \frac{\mu U}{\sigma_0}. \tag{7.2}$$

The capillary number is typically small so that an asymptotic approximation can be developed in the limit of $Ca \to 0$.

In Sect. 1.4, a two-dimensional lubrication-type model of evolution of a layer of initially uniform thickness d under the action of thermocapillary stresses was developed in the limit of small capillary numbers. In the present section, we consider the same configuration, shown in Fig. 7.1, except that the flow is driven by a nonuniform surfactant concentration rather than thermal gradients. Motivated by the same arguments as in Sect. 1.4, we choose $d/Ca^{1/3}$ and d as the scales for the Cartesian coordinates x and y shown in the sketch. The interface position is defined by a function $y = h(x,t)$, where t is the time variable scaled by $Ca^{-1/3}d/U$. The flow in the film is directed away from the region of higher surfactant concentration, which in Fig. 7.1 corresponds to the vicinity of $x = 0$. Note that here we omit tildas used to denote nondimensional variables in Chap. 1. Assuming gravity is negligible, the usual lubrication-type governing equations in a two-dimensional framework are written as

$$p_x = u_{yy}, \tag{7.3}$$

$$p_y = 0, \tag{7.4}$$

$$u_x + v_y = 0. \tag{7.5}$$

The velocity components u and v are scaled by U and $\mathrm{Ca}^{1/3}U$, respectively, and the pressure scale is $\mathrm{Ca}^{2/3}\sigma_0/d$. The leading-order interfacial boundary conditions in the limit of $\mathrm{Ca} \to 0$ are the same as for the two-dimensional lubrication-type models in the previous chapters, given by (2.6), (2.8), (2.9), and (2.10), except that the tangential stress condition has to include a contribution due to surfactants. Starting with a general dimensional version of this condition, (1.83), we observe that with our scales the leading-order term on the left-hand side of this equation is simply $\mu U u_y/d$. The derivation of this result follows the same steps as in the derivation of (1.88) in Sect. 1.5*. Let us introduce the nondimensional surfactant concentration defined by $\Gamma = \Gamma^*/\Gamma_0$, where Γ_0 is the maximum initial concentration. Using this definition and (7.1), the surface tension derivative term on the right-hand side of (1.83) is expressed as $\gamma_\Gamma \Gamma_0 \mathrm{Ca}^{1/3}\Gamma_x/d$, so that (1.83) becomes

$$\frac{\mu U}{d} u_y = -\frac{\gamma_\Gamma \Gamma_0 \mathrm{Ca}^{1/3}}{d} \Gamma_x. \tag{7.6}$$

The two sides of the equation are equally important in the limit of $\mathrm{Ca} \to 0$ only if the quantity $\mathrm{Ca}^{-2/3}\gamma_\Gamma \Gamma_0/\sigma_0$ is order one, which can be achieved simply by defining the velocity U from the condition

$$\left(\frac{\mu U}{\sigma_0}\right)^{2/3} = \frac{\gamma_\Gamma \Gamma_0}{\sigma_0}, \tag{7.7}$$

or, equivalently,

$$U = \frac{(\gamma_\Gamma \Gamma_0)^{3/2}}{\mu \sigma_0^{1/2}}. \tag{7.8}$$

The dimensionless tangential stress balance then takes the form

$$u_y = -\Gamma_x \quad \text{at} \quad y = h(x,t). \tag{7.9}$$

The definition of the characteristic velocity U above turns out to be most convenient for the simple problem discussed here, although other choices of U are often used in the literature, as discussed, e.g., in Craster and Matar [40].

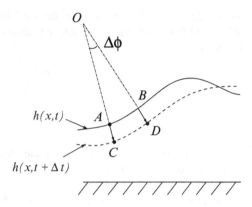

Fig. 7.2 Sketch for the derivation of surfactant transport equation, with the *solid line* showing interface shape at a time t and the *dashed line* at a time $t + \Delta t$

Following the usual steps of the lubrication-type analysis, described in detail, e.g., in Sects. 1.4 and 2.1, we find the velocity profile in the form

$$u = -\frac{1}{2}h_{xxx}\left(y^2 - 2yh\right) - \Gamma_x y, \tag{7.10}$$

and, by substituting this profile into the standard integral mass balance, (2.11), we obtain the evolution equation for the film thickness,

$$h_t + \left(\frac{h^3}{3}h_{xxx} - \frac{h^2}{2}\Gamma_x\right)_x = 0. \tag{7.11}$$

The surfactant concentration Γ in this equation is unknown and has to be determined from a separate equation. If the effects of chemical reactions and diffusion of surfactant along the interface are negligible, the equation for concentration is derived as follows. An arbitrary point on the interface can be defined by its horizontal coordinate x or by the arc length variable s defined as the distance from the point corresponding to $x = 0$, measured along the curve and scaled by $d/\text{Ca}^{1/3}$. For a positive x, the scaled arc length variable is related to the horizontal coordinate through

$$\frac{\mathrm{d}s}{\mathrm{d}x} = \sqrt{1 + \text{Ca}^{2/3}h_x^2}. \tag{7.12}$$

This formula implies that in the asymptotic limit of $\text{Ca} \to 0$ the difference between x and s is negligible so we can derive an equation for $\Gamma(s,t)$ and then simply replace s with x to obtain a formula for $\Gamma(x,t)$. A similar argument applies to the case of $x < 0$, leading to conclusion that s can be replaced with "$-x$" there.

Figure 7.2 shows snapshots of parts of the liquid–gas interface at a time t (*solid line*) and at a later time $t + \Delta t$ (*dashed line*). Consider an arbitrary point A on the solid line and a point B which is at a small distance Δs from A along the solid curve. The points C and D on the dashed line (showing the new interface shape) are chosen such that the straight line segments AC and BD are locally normal to the solid line;

then

$$|AC| = V_n^A \Delta t, \qquad |BD| = V_n^B \Delta t, \tag{7.13}$$

where V_n^A and V_n^B are the interface normal velocities at the points A and B, respectively. In the limit of $\Delta s \to 0$, the line segments AB and CD can be approximated by circular arcs of the radii κ^{-1} and $\kappa^{-1} + V_n \Delta t$, respectively, where κ is the curvature and V_n is the normal velocity, both at the point A; note that for small Δs we can use $V_n^B \approx V_n^A = V_n$. Both arcs are centered at a point O, shown in Fig. 7.2. If the angle between OA and OB is denoted by $\Delta \phi$, then $\Delta s = \kappa^{-1} \Delta \phi$ and the distance between C and D along the interface is found from

$$\Delta s' = \left(\kappa^{-1} + V_n \Delta t \right) \Delta \phi = \Delta s \left(1 + \kappa V_n \Delta t \right). \tag{7.14}$$

The local surfactant mass balance is written as

$$\Gamma_C \Delta s' - \Gamma_A \Delta s = (j_A - j_B) \Delta t, \tag{7.15}$$

where Γ_C and Γ_A are the concentrations, j_A and j_B are the surfactant fluxes at the respective points. We note that for finite Δs, variations of Γ along AB and CD would have to be taken into account when formulating the mass balance, but in the limit of $\Delta s \to 0$ such variations do not contribute to the leading-order mass balance. Similarly, variations of the fluxes along AC and BD can be neglected in the limit of $\Delta t \to 0$. For negligible diffusion, the fluxes j_A and j_B are determined by the amount of surfactant carried by the flow, so they can be expressed in terms of the values of the velocity along the interface, $\hat{u}_A$ and $\hat{u}_B$, as follows:

$$j_A = \Gamma_A \hat{u}_A, \qquad j_B = \Gamma_B \hat{u}_B. \tag{7.16}$$

Substituting (7.14) and (7.16) into the surfactant mass balance condition given by (7.15), we obtain

$$(\Gamma_C - \Gamma_A) \Delta s + \Gamma_C \kappa V_n \Delta t \Delta s = (\Gamma_A \hat{u}_A - \Gamma_B \hat{u}_B) \Delta t. \tag{7.17}$$

Taking the limit of $\Delta t \to 0$, $\Delta s \to 0$ and assuming that the arc length value corresponding to the point A is equal to s, we find

$$[\Gamma_t]_n + \Gamma(s,t) \kappa(s,t) V_n(s,t) + (\Gamma(s,t) \hat{u}(s,t))_s = 0. \tag{7.18}$$

Here we introduced the time derivative of Γ in the direction normal to the interface [32, 141] at the point A, defined by

$$[\Gamma_t]_n = \lim_{\Delta t \to 0} \frac{\Gamma_C - \Gamma_A}{\Delta t}. \tag{7.19}$$

In the lubrication-type approximation considered in the present section, the variable s can be replaced by x based on (7.12), as discussed above, the term proportional to the normal velocity V_n is asymptotically negligible, and the derivative $[\Gamma_t]_n$ is the same as the partial derivative of $\Gamma(x,t)$ with respect to t. The velocity along the interface, $\hat{u}$, can be found from (7.10) by evaluating the horizontal velocity component u at the interface, $y = h(x,t)$. The equation for surfactant concentration, $\Gamma(x,t)$, then takes the form

$$\Gamma_t + \left[\left(\frac{1}{2}h^2 h_{xxx} - h\Gamma_x \right) \Gamma \right]_x = 0. \tag{7.20}$$

The effect of surfactant diffusion can be incorporated by adding the diffusion fluxes, assumed proportional to the local derivative of concentration, to the surfactant mass balance. The resulting equation is then

$$\Gamma_t - \mathrm{Pe}_s^{-1}\Gamma_{xx} + \left[\left(\frac{1}{2}h^2 h_{xxx} - h\Gamma_x \right) \Gamma \right]_x = 0, \tag{7.21}$$

where Pe_s is the surface Peclet number, defined in terms of the surfactant diffusion coefficient D_s according to

$$\mathrm{Pe}_s = \frac{Ud}{D_s \mathrm{Ca}^{1/3}}. \tag{7.22}$$

In most practical situations, the surface Peclet number is large (above 100). By simultaneously solving (7.11) and (7.21) with appropriate initial and boundary conditions, one can determine the time-dependent interface shape and the distribution of the surfactant. It is important to emphasize that these two quantities are coupled together: the right-hand-side of (7.21) depends on the interface shape, $h(x,t)$, while (7.11) depends on $\Gamma(x,t)$.

The extent of the film in the configuration shown in Fig. 7.1 is assumed infinite, but the numerical solution has to be carried out on a finite domain. To illustrate the numerical technique, we consider a special case when the interface is symmetric with respect to the y-axis so that an interval $[0,L]$ can be chosen as the computational domain. The length L has to be sufficiently large so that the variations of both $h(x,t)$ and $\Gamma(x,t)$ are negligible for $|x| > L$. The boundary conditions at $x = L$ are then formulated as

$$h(L,t) = 1, \quad h_x(L,t) = 0, \quad \Gamma(L,t) = \Gamma_x(L,t) = 0. \tag{7.23}$$

The symmetry conditions at $x = 0$ are in the form

$$h_x(0,t) = h_{xxx}(0,t) = \Gamma_x(0,t) = \Gamma_{xxx}(0,t) = 0. \tag{7.24}$$

Initially, the film thickness is uniform and the surfactant distribution is taken to be Gaussian to simulate a situation when the surfactant is initially localized near $x = 0$:

$$h(x,0) = 1, \quad \Gamma(x,0) = \mathrm{e}^{-\alpha x^2}. \tag{7.25}$$

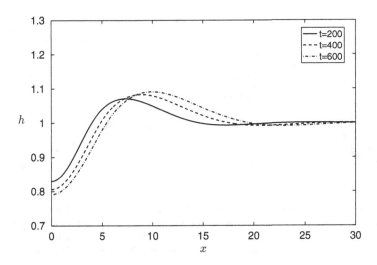

Fig. 7.3 Interface shapes at different times found from the numerical solution of the system of (7.11) and (7.21) for $\text{Pe}_s = 10^3$

The numerical finite difference solution of (7.11) and (7.21) with the conditions (7.23)–(7.25) is then carried out using essentially the same approach as was described in detail in the study of the thermocapillary flow in Sect. 1.4. The MATLAB code for this problem can be found in the appendix, Sect. B.7. Typical results for the interface shape at different times are shown in Fig. 7.3. Since the surfactant concentration is higher near $x = 0$, the surface tension is lower there and the liquid flows away from this region, resulting in the formation of a depression near $x = 0$, as seen in the figure. This dynamics is qualitatively similar to the case of thermocapillary flow from Sect. 1.4, but there are also important differences due to the fact that the surface tension gradient is changing significantly as the surfactant spreads on the surface of the film. As a result, the extent of the region of high-interface deformation is significantly wider than the length scale of the initial concentration profile, $\sim \alpha^{-1/2} = 0.1$. Note that for the thermocapillary flow results shown in Fig. 1.7 the interface was deformed only in the region of the order of the length scale of the localized temperature profile, $\sim \tilde{\alpha}^{-1/2}$.

The dynamics of surfactant spreading and corresponding interface deformation over several orders of magnitude in the nondimensional time t has been investigated by Jensen and Grotberg [73] based on the numerical solution of the system of equations

$$h_t + \left(\frac{C_m h^3}{3} h_{xxx} - \frac{h^2}{2} \Gamma_x \right)_x = 0, \qquad (7.26)$$

$$\Gamma_t - \text{Pe}_s^{-1} \Gamma_{xx} + \left(\frac{C_m}{2} h^2 \Gamma h_{xxx} - h \Gamma \Gamma_x \right)_x = 0. \qquad (7.27)$$

These equations are derived using essentially the same lubrication-type approach as in the derivation of (7.11)–(7.21) above, but the horizontal length scale L_x in [73] is

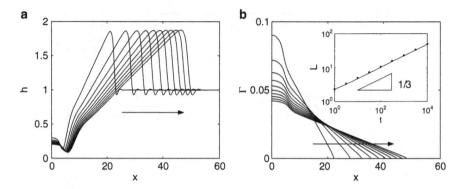

Fig. 7.4 Film thickness and concentration profiles at different times found from the numerical solution of the coupled system of equations for concentration and film thickness of Jensen and Grotberg [73] for $Pe_s = 10^6$, $C_m = 10^{-3}$. The inset shows the position of the leading edge of the spreading surfactant. Reprinted from [40] with permission of the American Physical Society

defined by the initial extent of the surfactant monolayer rather than $d/Ca^{1/3}$. As a result, an additional parameter

$$C_m = \frac{d^2(\sigma_0 - \gamma_\Gamma \Gamma_0)}{L_x^2 \gamma_\Gamma \Gamma_0} \tag{7.28}$$

appears in the system of equations for film thickness and surfactant concentration. The boundary conditions (7.23)–(7.24) together with the initial conditions

$$h(x,0) = 1, \quad \Gamma(x,0) = \frac{1}{2}(1 + \tanh(100[1-x])) \tag{7.29}$$

were used to obtain the numerical solutions of (7.26)–(7.27) shown in Fig. 7.4 (reprinted from [40]). The results for both thickness and surfactant concentration at different moments of time are shown in Fig. 7.4 for $Pe_s = 10^6$. A sharp front in the deforming interface profile is seen near the point separating surfactant-covered and surfactant free ($\Gamma = 0$) parts of the interface, with the film thickness changing by approximately a factor of two in the relatively narrow range of values of x.

The rate at which the surfactant spreads over the surface of the film can be estimated based on the following considerations. Suppose at a time t, the extent of the surfactant coverage of the interface is characterized by the nondimensional length $L = L(t)$ and the average (over x) interface height by $\bar{h} = \bar{h}(t)$. Surfactant concentration can then be estimated as M/L, where M is the total mass of the surfactant, expressed in nondimensional variables. The evolution equation (7.26) for film thickness implies that the following estimates can be made,

$$\frac{\bar{h}}{t} \sim \frac{\bar{h}^2 M/L}{L^2}. \tag{7.30}$$

Note that since the time-dependent interface deformation is caused by surfactant concentration gradients, the first and the last terms in (7.26) have to be comparable, leading to (7.30). Assuming that $\bar{h} \sim 1$ throughout the spreading process, (7.30) leads to

$$L \sim t^{1/3}. \tag{7.31}$$

Thus, the characteristic size of the region covered by surfactant grows with time as $t^{1/3}$, suggesting a possibility of solutions such that the film thickness depends only on $\xi = x/t^{1/3}$. These special solutions are referred to as the similarity solutions, following the general terminology introduced in the discussion of film rupture in Sect. 3.5.3. In the present context, a similarity solution is of the form

$$h(x,t) = H(\xi), \quad \Gamma(x,t) = t^{-1/3}G(\xi), \tag{7.32}$$

where the functions $H(\xi)$ and $G(\xi)$ are to be determined, the factor of $t^{-1/3}$ in the expression for concentration is due to the condition of conservation of mass ($M = $ const.). Substituting (7.32) into (7.26)–(7.27) and taking the limit of small $\mathrm{Pe_s}^{-1}$ and C_m, we obtain

$$\frac{1}{3}H'\xi + \frac{1}{2}\left(H^2 G'\right)' = 0, \tag{7.33}$$

$$\frac{1}{3}(G\xi)' + (HGG')' = 0, \tag{7.34}$$

where primes denote derivatives with respect to ξ. The second equation can be integrated, resulting in $(\xi + 3HG')\, G = c_1$. The constant of integration, c_1, has to be zero to satisfy the condition of $G = 0$ in the region not yet covered by the surfactant, leading to

$$G' = -\frac{\xi}{3H}. \tag{7.35}$$

This relation allows us to rewrite (7.33) as a separable ordinary differential equation for $H(\xi)$, with the general solution in the form $H(\xi) = c_2 \xi$, $c_2 = $ const. Substituting this solution into (7.35) leads to

$$G = -\frac{\xi}{3c_2} + c_3, \quad c_3 = \text{const.} \tag{7.36}$$

Motivated by the presence of a sharp front in the numerical interface profiles shown in Fig. 7.4a, we seek a solution such that the functions H and G' are discontinuous at a single location in rescaled coordinates, $\xi = \xi_s$. Integrating (7.33) from $\xi_s - \Delta\xi$ to $\xi_s + \Delta\xi$ and taking the limit of $\Delta\xi \to 0$ leads to the following condition upstream from the front,

$$\frac{1}{3}\left(1 - H\left(\xi_s^-\right)\right)\xi_s - \frac{1}{2}\left(H\left(\xi_s^-\right)\right)^2 G'\left(\xi_s^-\right) = 0. \tag{7.37}$$

Here, we found the integral of $H'\xi$ by expressing the singular part of the derivative of the discontinuous function $H(\xi)$ as $(1 - H(\xi_s^-))\,\delta(\xi - \xi_s)$, where $\delta(\xi - \xi_s)$ is the delta function. Substituting $H(\xi) = c_2\xi$ together with (7.35) into (7.37), we find that $c_2 = 2\xi_s^{-1}$ and therefore

$$H(\xi) = \frac{2\xi}{\xi_s}. \tag{7.38}$$

Using $c_2 = 2\xi_s^{-1}$ together with the condition $G(\xi_s) = 0$ in (7.36) leads to

$$G(\xi) = \frac{1}{6}\xi_s(\xi_s - \xi). \tag{7.39}$$

We note that the possibility of solutions with $G(\xi_s) > 0$ is not considered here since they have not been observed in any of the numerical simulations, an indication that they are not likely to be stable. Finally, to define the location of the front we use the condition of conservation of the total nondimensional surfactant mass, M. Integrating the initial concentration profile (7.29) over the entire real axis gives $M = 2$, so the mass conservation condition can be written as

$$\int_{-\infty}^{\infty} \Gamma(x,t)\,\mathrm{d}x = 2 \tag{7.40}$$

or, equivalently, as

$$\int_{0}^{\xi_s} G(\xi)\,\mathrm{d}\xi = 1. \tag{7.41}$$

Substituting (7.39) into this integral condition gives $\xi_s^3 = 12$, completing our derivation of the similarity solutions for thickness and concentration profiles. The final result can be expressed in terms of the original variables as

$$h(x,t) = \frac{2x}{(12t)^{1/3}}, \quad \Gamma(x,t) = \frac{2}{(12t)^{1/3}}\left[1 - \frac{x}{(12t)^{1/3}}\right], \quad x \le L \tag{7.42}$$

and $h(x,t) = 1$, $\Gamma(x,t) = 0$ for $x > L$. Here, the position of the front $L = L(t)$ is given by

$$L = (12t)^{1/3}. \tag{7.43}$$

The similarity solutions turn out to be in good agreement with the numerical results obtained from the full time-dependent system of (7.26)–(7.27) for small C_m and Pe_s^{-1} and at sufficiently large times, as discussed in detail, e.g., in [40]. To illustrate the agreement, the inset in Fig. 7.4b shows the position of the surfactant leading edge predicted by (7.43), solid line, and obtained from the numerical solution, shown by the dots. Both are plotted in log–log coordinates, so that the prediction of (7.43) is a straight line of slope 1/3.

The two-dimensional model of the present section is useful for illustrating the key physical effects in surfactant spreading, but in experiments an axisymmetric configuration is easier to realize. Jensen and Grotberg [73] extended the lubrication-type model of surfactant spreading on initially uniform film to the axisymmetric geometry. Their numerical solutions indicate that the characteristic radius of the area covered by an initially localized surfactant expands in time as $t^{1/4}$. This can be explained by the following argument. The condition of conservation of mass in the axisymmetric geometry leads to a relation between the characteristic values of the concentration $\bar{\Gamma}$ and the radius of the circular area covered by surfactant R_s in the form $\bar{\Gamma} R_s^2 = \mathrm{const}$. The axisymmetric version of (7.26) leads to an estimate,

$$\frac{\bar{h}}{\bar{t}} \sim \frac{\bar{h}^2 \bar{\Gamma}}{R_s^2},\tag{7.44}$$

which together with $\bar{h} \sim 1$ gives $R_s \sim t^{1/4}$, consistent with numerical simulations. Similarity solutions for interface shape and surfactant concentration can be obtained for the case of axisymmetric geometry, as discussed in detail in Jensen and Grotberg [73].

Estimating the rate of surfactant spreading is important for many practical applications. For example, surfactants are essential for normal functioning of human lungs. In prematurely born infants, lack of surfactants often leads to severe medical problems, a condition known as infant respiratory distress syndrome (IRDS). A promising treatment for IRDS involves injection of artificial surfactant to compensate for the deficiency. Better understanding of how surfactants spreads on liquid layers is essential for improving the effectiveness of this treatment.

7.2* General Equation for Interfacial Surfactant Transport

The two-dimensional framework used in the previous section provides important insights into the nature of physical phenomena associated with surfactants, but is not directly applicable to describing experimental configurations which are always three-dimensional, with two-dimensional interfaces separating different phases. A derivation of the transport equation for insoluble surfactant at a two-dimensional liquid–gas interface was provided by Stone [124]. His approach is based on the ideas of material line and surface elements. Prior to discussing the derivation of the equation for Γ in three-dimensional configurations, we illustrate how the idea of material line elements can be applied to the two-dimensional configuration of the previous section, providing a slightly different derivation of the equation for Γ. The sketch of the geometric configuration with interface shapes at two different times separated by Δt is shown in Fig. 7.5. We use the same nondimensional variables as in the previous section and neglect the effects of diffusion. Suppose the points A and B mark the locations of two fluid elements adjacent to the interface at a time t. As is common in fluid mechanics [1, 12], a fluid element is defined as a region of fluid with dimensions much smaller than the scale of the flow but still containing a very large number of molecules so that the continuum description

Fig. 7.5 Sketch for the
alternative derivation of
surfactant transport equation,
with A' and B' showing the
new locations of the fluid
elements initially at the points
A and B

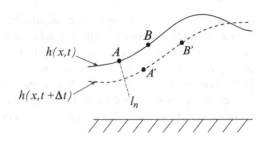

is valid. As before, the distance between the points A and B measured along the interface is denoted by Δs and the value of the arc length variable corresponding to the point A by s. As the interface deforms, the two fluid elements move, ending up at the points A' and B' at the time $t + \Delta t$. Suppose $\hat{s}$ is the arc length variable measured from the point A. Then, if the scaled flow velocity along the interface is $\hat{u}(\hat{s}, t)$, the value of $\hat{s}$ corresponding to the point A' is

$$\hat{s} = \hat{u}(0, t)\Delta t, \tag{7.45}$$

while the new distance between the fluid elements is

$$\Delta s' = \Delta s + \hat{u}(\Delta s, t)\Delta t - \hat{u}(0, t)\Delta t + \kappa V_n \Delta t \Delta s. \tag{7.46}$$

The last term in the equation, derived using the same approach as in the previous section, accounts for the fact that even when the fluid elements do not move along the interface, the distance between them can change because of interface stretching/compression; κ and V_n are the curvature and the normal velocity, respectively, evaluated at the point A. It is important to keep the curvature-dependent term in the equation since for small Δt and Δs this term is of the same order of magnitude as the correction due to velocity difference between the two fluid elements.

The condition of conservation of the amount of surfactant between the two fluid elements is written as

$$\Gamma(0, t)\Delta s = \Gamma\left(\hat{u}(0, t)\Delta t, t + \Delta t\right)\Delta s'. \tag{7.47}$$

Substituting (7.46) into this condition and using the Taylor expansions valid for small Δt, we obtain

$$\Gamma_{\hat{s}}(0, t)\hat{u}(0, t)\Delta s + \Gamma_t(0, t)\Delta s + \Gamma(0, t)\left(\hat{u}(\Delta s, t) - \hat{u}(0, t) + \kappa V_n \Delta s\right) = 0. \tag{7.48}$$

Dividing by Δs and taking the limit of $\Delta s \to 0$ leads to

$$\Gamma_t(0, t) + \left(\Gamma(\hat{s}, t)\,\hat{u}(\hat{s}, t)\right)_{\hat{s}}\big|_{\hat{s}=0} + \Gamma(0, t)\kappa V_n = 0. \tag{7.49}$$

The relationship between the variable $\hat{s}$ and the original arc length variable s (measured from the point at the interface corresponding to $x = 0$) is in general rather

Fig. 7.6 Sketch for the general derivation of the surfactant transport equation using material derivatives

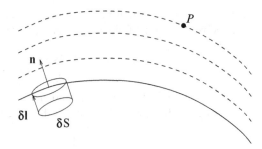

complicated and depends on the shape of the entire interface between $s = 0$ and the value of s corresponding to the point A. However, the equation for Γ only involves instantaneous derivatives with respect to $\hat{s}$ at the point A, and these derivatives are the same as partial derivatives with respect to s, resulting in

$$[\Gamma_t]_n + (\Gamma(s,t)\hat{u}(s,t))_s + \Gamma(s,t)\kappa(s,t)V_n(s,t) = 0, \qquad (7.50)$$

which is the same equation for surfactant concentration as in the previous section.

The same approach can be extended to three dimensions and expressed in more compact notation following Stone [124]. Here we present a slightly modified version of his derivation, taking into account clarifications from the subsequent studies [32, 141]. The derivation is not restricted to a thin-film geometry, so all velocity components are now scaled by the same characteristic velocity U, and the liquid is not necessarily incompressible. Consider a portion of the interface, shown schematically by the curved solid line in Fig. 7.6. The dashed lines show segments of the interface at later times. Then, for an arbitrary point P ahead of the advancing interface, but sufficiently close to it, one can define a new function $\tilde{\Gamma}(x,y,z)$ by recording the value of the surfactant concentration Γ at the point P at the moment in time when the advancing interface passes through P. This approach only considers a small advance in the interface toward the point P to define the function $\tilde{\Gamma}$ locally so that the definition is locally unique. By construction, $\tilde{\Gamma}$ is not a function of time, so $\partial\tilde{\Gamma}/\partial t = 0$ and therefore the material derivative of $\tilde{\Gamma}$ can be written as

$$\frac{D\tilde{\Gamma}}{Dt} = \mathbf{u} \cdot \nabla\tilde{\Gamma}. \qquad (7.51)$$

Consider a fluid element of cylindrical shape, with the top being a patch of the interface of the area $\delta S(t)$ and the side surface generated by the vector $\delta\mathbf{l}$, as shown in Fig. 7.6. Note that a fluid element adjacent to the interface will not move away from it into the bulk since the normal velocity of fluid at the interface is always the same as the normal velocity of the interface itself. The condition of conservation of the amount of surfactant ($\tilde{\Gamma}\delta S$) in the absence of diffusion can then be expressed as

$$\frac{D\tilde{\Gamma}}{Dt}\delta S + \tilde{\Gamma}\frac{d\delta S}{dt} = 0. \qquad (7.52)$$

If **n** the local unit normal vector to the interface, shown in the sketch in Fig. 7.6, then the volume of the fluid element is expressed as $\delta V = \delta S \mathbf{n} \cdot \delta \mathbf{l}$ and therefore its rate of change is

$$\frac{d\delta V}{dt} = \frac{d(\delta S \mathbf{n})}{dt} \cdot \delta \mathbf{l} + \delta S \mathbf{n} \cdot \frac{d\delta \mathbf{l}}{dt}. \tag{7.53}$$

The rate of change of volume of a fluid element is expressed in terms of the velocity field using the divergence theorem (as shown, e.g., in Batchelor [12, p. 75]),

$$\frac{d\delta V}{dt} = \delta V \, \nabla \cdot \mathbf{u}, \tag{7.54}$$

while the rate of change for a line segment $\delta \mathbf{l}$ moving with the flow is expressed in terms of the difference between the velocities at the two ends, leading to

$$\frac{d\delta \mathbf{l}}{dt} = \delta \mathbf{l} \cdot \nabla \mathbf{u}. \tag{7.55}$$

Therefore, (7.53) can be written as

$$\delta V \, \nabla \cdot \mathbf{u} = \frac{d(\delta S \mathbf{n})}{dt} \cdot \delta \mathbf{l} + \delta S \mathbf{n} \cdot (\delta \mathbf{l} \cdot \nabla \mathbf{u}), \tag{7.56}$$

valid for an arbitrary $\delta \mathbf{l}$. By choosing $\delta \mathbf{l} = \mathbf{n}\delta l$ and observing that

$$\mathbf{n} \cdot \frac{d\mathbf{n}}{dt} = \frac{1}{2} \frac{d(\mathbf{n} \cdot \mathbf{n})}{dt} = 0, \tag{7.57}$$

we obtain

$$\delta S \nabla \cdot \mathbf{u} = \frac{d\delta S}{dt} + \delta S \mathbf{n} \cdot (\mathbf{n} \cdot \nabla \mathbf{u}). \tag{7.58}$$

From this equation, the rate of change of δS can be expressed in terms of the surface gradient operator $\nabla_s = (\mathbf{I} - \mathbf{nn}) \cdot \nabla$ as follows,

$$\frac{d\delta S}{dt} = \delta S \nabla_s \cdot \mathbf{u}. \tag{7.59}$$

Substituting this formula together with (7.51) into (7.52), we arrive at

$$\mathbf{u} \cdot \nabla \tilde{\Gamma} + \tilde{\Gamma} \nabla_s \cdot \mathbf{u} = 0. \tag{7.60}$$

Alternatively, this expression can be written as

$$\mathbf{u} \cdot \mathbf{nn} \cdot \nabla \tilde{\Gamma} + \nabla_s \cdot (\tilde{\Gamma} \mathbf{u}) = 0. \tag{7.61}$$

The first term on the left-hand side represents the derivative of $\tilde{\Gamma}$ in the direction normal to the interface, which is the same as the time derivative of Γ taken along the normal as the interface advances; the latter quantity is denoted by $[\partial\Gamma/\partial t]_n$. We also observe that the surface gradient operator acts along the surface, so that $\tilde{\Gamma}$ in (7.61) can be replaced by Γ, resulting in

$$\left[\frac{\partial\Gamma}{\partial t}\right]_n + \nabla_s \cdot (\Gamma\mathbf{u}) = 0. \tag{7.62}$$

Finally, the velocity $\mathbf{u}$ can be decomposed into components along the interface and normal to it, leading to the equation

$$\left[\frac{\partial\Gamma}{\partial t}\right]_n + \nabla_s \cdot (\Gamma\mathbf{u}_s) + \Gamma(\nabla_s \cdot \mathbf{u})(\mathbf{u} \cdot \mathbf{n}) = 0, \tag{7.63}$$

where $\mathbf{u}_s$ denoted the velocity component along the interface; we also used the fact that $\nabla_s\Gamma$ is orthogonal to $\mathbf{n}$.

For the special case of a stationary interface, the condition of the conservation of the amount of surfactant in a control volume adjacent to the interface leads to the simple equation

$$\nabla_s \cdot (\Gamma\mathbf{u}_s) = 0. \tag{7.64}$$

The effect of diffusion can then be included in (7.63) by adding the diffusive fluxes at the boundary of the surface element, resulting in

$$\left[\frac{\partial\Gamma}{\partial t}\right]_n + \nabla_s \cdot (\Gamma\mathbf{u}_s) + \Gamma(\nabla_s \cdot \mathbf{u})(\mathbf{u} \cdot \mathbf{n}) = D_s\nabla_s^2\Gamma, \tag{7.65}$$

where D_s is the surface diffusion coefficient. Equation (7.65) can also be derived by generalizing the approach of Sect. 7.1 to three-dimensional configurations, by considering patches of a two-dimensional interface instead of line segments such as AB and CD shown in Fig. 7.2. The details of this derivation are presented in Wong et al. [141].

7.3 Models of Soluble Surfactants

The model of insoluble surfactant discussed in the previous two sections is based on the assumption that there is no surfactant in the bulk and therefore the transport of the surfactant in the system is described by a single equation for the interfacial surfactant concentration Γ. Most surfactants are soluble to some extent, so an additional quantity, the dimensional bulk concentration c^*, has to be introduced to describe their distribution in the bulk of the liquid, i.e., away from the interface, and their interchange between bulk and interface.

For surfactant molecules in the bulk, the concentration c^* satisfies the convection–diffusion equation, as would be the case for any dissolved chemical. However, there is a thin layer of characteristic thickness l near the interface, in which the tendency of surfactant molecules to accumulate at the interface affects their dynamics. For a surfactant molecule in this layer, there is a significant probability of ending up at the interface, i.e., adsorption. Clearly, the flux of surfactant molecules j_a from the interfacial layer to the interface is proportional to their local concentration in the interfacial layer, assuming the probability of adsorption for each molecule is fixed. The constant of proportionality is referred to as an adsorption constant and denoted by k_a. The overall tendency of surfactant molecules to accumulate at interface does not mean that individual molecules cannot leave the interface and end up in the bulk (a process called desorption) in the course of thermal motion. The probability of this event is characterized by a desorption constant k_d, with desorption flux j_d proportional to the interfacial concentration so that $j_d = k_d \Gamma^*$. Equilibrium is reached when the two fluxes are equal so that

$$k_d \Gamma^* = k_a c^*. \tag{7.66}$$

The thickness l of the interfacial layer is very small compared to other dimensions of the system, so the standard approach to modeling soluble surfactant involves solving the diffusion equation up to the interface (instead of solving it up to the surface located at a distance l below the interface) and then equating the difference $j_a - j_d$ to the local diffusion flux at the interface, $-D_b(\mathbf{n} \cdot \nabla)c^*$, where D_b is the bulk diffusion constant.

To illustrate the basic ideas of modeling of soluble surfactants, we use the same two-dimensional configuration as in Sect. 7.1 shown in Fig. 7.1. The dimensional equilibrium condition (7.66) suggests a natural definition of the nondimensional concentration as

$$c = \frac{k_a c^*}{k_d \Gamma_0}. \tag{7.67}$$

The scaled flux J of molecules between the interface and the bulk,

$$J = \frac{j_a - j_d}{\Gamma_0 U / L_x}, \tag{7.68}$$

is then expressed in terms of scaled concentrations as

$$J = K(c_s - \Gamma), \tag{7.69}$$

where $K = k_d L_x / U$ is the ratio of timescales of the flow and the desorption process, c_s is the scaled concentration at the surface of the liquid film. The choice of horizontal length and velocity scales, L_x and U, as well as other scales used in the problem, is discussed in detail in [40,73,74]. With these choices, the equation for the

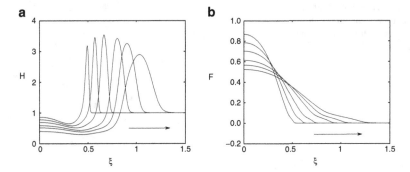

Fig. 7.7 Interface shapes at different times found from the numerical solution of the system of equations describing spreading of soluble surfactants on the surface of initially uniform liquid film. Reprinted from [74] with permission of American Institute of Physics

scaled film thickness, (7.26), is equally applicable for the case of soluble surfactant, while the equation for surfactant concentration has to be modified to account for the surfactant flux J into the bulk, leading to

$$\Gamma_t = \mathrm{Pe_s}^{-1}\Gamma_{xx} - \left(\frac{C_m}{2}h^2\Gamma h_{xxx} - h\Gamma\Gamma_x\right)_x + J. \qquad (7.70)$$

Application of the lubrication-type scalings shows that at leading order the concentration c is a function of x and t only and satisfies the equation

$$c_t = \frac{1}{h\mathrm{Pe_b}}(hc_x)_x - \left(\frac{C_mh^2}{3}h_{xxx} - \frac{h}{2}\Gamma_x\right)c_x - \frac{\beta J}{h}, \qquad (7.71)$$

where $\mathrm{Pe_b}$ is the bulk Peclet number, $\beta = k_a/dk_d$ is the solubility of the surfactant. The flux J given by (7.69) has to be equal to the scaled diffusion flux near the interface, leading to the interfacial boundary condition

$$J = \frac{L_x^2}{\beta \mathrm{Pe_b} d^2}c_y. \qquad (7.72)$$

Numerical simulations of the system of (7.11), (7.70), and (7.71) for different parameter values have been conducted by several authors, as discussed in [40]. Due to effects of solubility, the shape of the advancing region of the high interface deformation tends to be a narrow pulse rather than a front observed for insoluble surfactant. This is illustrated by the plot of interface shape in Fig. 7.7a, obtain by Jensen and Grotberg [74] in the limit of $C_m = 0$, $K \to \infty$. As in the previous section, the variable ξ defined by $\xi = xt^{-1/3}$. Corresponding surfactant distributions, represented by the function $F = ct^{1/3}$, are shown in Fig. 7.7b.

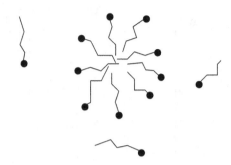

Fig. 7.8 A sketch illustrating formation of micelles in a surfactant solution above the critical micelle concentration

7.4   High Concentration of Surfactants

The structure of typical surfactant molecules, i.e., the presence of hydrophobic tails and hydrophilic heads, can also affect their behavior in the bulk of the liquid as the concentration c^* increases. The previously discussed models assumed that surfactant is dilute so that surfactant molecules make up a small fraction of the aqueous solution and their behavior is governed by the diffusion equation. For sufficiently large concentrations, it is energetically favorable for the molecules to aggregate in bulk, forming structures called micelles, as sketched in Fig. 7.8 for a polar liquid such as water. The hydrophilic heads of the molecules are in contact with the aqueous solution outside the micelle while the hydrophobic tails are inside the aggregate. The physical reason why such arrangement is advantageous has to do with reducing the contact between the hydrophobic tails and water, i.e., is essentially the same as the reason why surfactant molecules tend to adsorb at the water–air interface with the hydrophobic tails sticking out of the liquid phase. Other geometric shapes of micelles are possible, e.g., cylindrical and elliptic. Also, the so-called reverse micelles with the tails inside the aggregate have been observed in nonpolar liquids. The minimum concentration at which micelles form is called the critical micelle concentration and can be determined by comparing the free energies of the system with and without the micelles.

Formation of micelles plays an important role in the cleaning action of detergents. Tiny oil and grease droplets often end up inside the micelles are thus are separated from the fabric and easily washed away. This is only one aspect of how detergents work. The reduction of the surface tension due to action of surfactants also plays an important role regardless of whether the conditions are favorable for formation of micelles.

The effects due to formation of micelles can be incorporated into lubrication-type mathematical models discussed in the previous sections by introducing a new nondimensional variable m describing the micelle concentration, as discussed in [40]. The system is then described by four partial differential equations for the four unknown functions: the liquid layer thickness h, the interfacial and bulk concentrations of surfactant molecules which are not parts of micelles, Γ and c, and the micelle concentration m. An example of numerical solution of such system is discussed in Edmonstone et al. [48] in the context of spreading of a thin droplet of liquid containing a soluble surfactant above the critical micelle concentration. They found that under these condition there is a distinct protuberance formed at the leading edge of the spreading droplet, similar in shape to the capillary ridge seen in the gravity-driven flow discussed in Chap. 2. This protuberance can separate from the main droplet for sufficiently high values of the total mass of the surfactant. Edmonstone et al. [48] also discuss stability criteria for spreading process and show that a fingering instability qualitatively similar to the one discussed in Chap. 2 can develop at the leading edge of the droplet. These conclusions are in qualitative agreement with experimental observations. Numerical simulations of surfactant-driven fingering instability can be conducted using the approaches similar to the one discussed in Chap. 2 in the context of gravity-driven flows.

7.5   Notes on Literature

Models of thin liquid films containing surfactants are developed by Jensen and Grotberg [73,74] and reviewed by, e.g., Oron et al. [97] and Craster and Matar [40]. Due to many practical applications of surfactants, most notably as soaps and detergents, their classifications and properties are discussed in a number of books which we do not attempt to list here. The key physical phenomena important for the development of mathematical models of flows with surfactants are described, e.g., in Israelachvili [72], Adamson and Gast [2], and Hunter [62].

Appendix A
Selected Definitions and Theorems
from Thermodynamics

A thermodynamic system is *closed* when it can exchange energy but not matter with its exterior. For a closed system maintained at constant temperature T and volume V, the Helmholtz free energy,

$$F = U - TS, \tag{A.1}$$

is minimized at equilibrium. Here, U is the total internal energy, S is the entropy of the system. For a closed system at constant temperature T and pressure p, the Gibbs free energy,

$$G = U + PV - TS, \tag{A.2}$$

is minimized. The differentials of $F(V,T)$ and $G(P,T)$ can be expressed as:

$$dF = -pdV - SdT, \quad dG = Vdp - SdT. \tag{A.3}$$

Electrochemical potential $\bar{\mu}_i$ includes the effects of electric fields and is defined by:

$$\bar{\mu}_i = \mu_i + zeN_A\psi, \tag{A.4}$$

where the potential of the electrical field is ψ; μ_i is the chemical potential that a charged species would have if it were to suddenly loose its electric charge.

For homogeneous systems containing different species, numbered by an integer k, the differential of the energy is determined by:

$$dU = TdS - PdV + \Sigma\mu_k dN_k, \tag{A.5}$$

where μ_k are the chemical potentials. For the Gibbs and Helmholtz free energies, the following equations can then be used,

$$dG = Vdp - SdT + \Sigma\mu_k dN_k, \quad dF = -pdV - SdT + \Sigma\mu_k dN_k. \tag{A.6}$$

The same equations have been applied to heterogeneous systems containing different species/phases when the effects of surface tension are negligible. In

V.S. Ajaev, *Interfacial Fluid Mechanics: A Mathematical Modeling Approach*, DOI 10.1007/978-1-4614-1341-7, © Springer Science+Business Media, LLC 2012

the studies incorporating the effects of surface tension, the Gibbs free energy of a heterogeneous system is assumed to be a sum of the Gibbs free energies of individual phases and the contributions from the interfaces. Each of the latter is proportional to the area of an interface, with the coefficient of proportionality being the surface tension.

Appendix B
MATLAB Codes

B.1 Thermocapillary flow

```
function tc
global N L alpha
alpha = 1;

% spatial mesh, initial condition

N = 4000;
L = 20;
x = (L/N)*(0:N-1);
y0 = ones(1,N);

% sparsity matrix for the Jacobian

e = ones(N,1);
S = spdiags([e e e e e], -2:2,N,N);
t = 0:3:9;
options = odeset('RelTol',1e-4,'AbsTol',1e-20, ...
'JPattern',S,'BDF','on');
[t,h] = ode15s(@f,t,y0,options);
plot(x,h(2,:),'Color','black','LineWidth',2)
hold on
plot(x,h(3,:),'--','Color','black','LineWidth',2)
plot(x,h(4,:),'-.', 'Color','black','LineWidth',2)
xlabel('$\tilde{x}$','Interpreter','Latex','Fontsize',20)
ylabel('$\tilde{h}$','Interpreter', ...
'Latex','Fontsize',20,'Rotation',0)
legend('t=3','t=6','t=9')
axis([0 12 0 1.4])

%-------------------------------------------------------
```

V.S. Ajaev, *Interfacial Fluid Mechanics: A Mathematical Modeling Approach*,
DOI 10.1007/978-1-4614-1341-7, © Springer Science+Business Media, LLC 2012

```matlab
function ht = f(t,y)
global N L alpha
dx = L/N;

y(N+1) = 1;
y(N+2) = y(N);
for i = 3:N
    xp = L*(i-0.5)/N;
    yx3p = (y(i+2) - 3*y(i+1) + 3*y(i) - y(i-1)) / dx^3;
    QP=((y(i+1)^3 + y(i)^3)/6)*yx3p + ((y(i+1)^2 + ...
    y(i)^2)/2)*alpha*xp*exp(-alpha*xp^2);
    xm = L*(i-1.5)/N;
    yx3m = (y(i+1) - 3*y(i) + 3*y(i-1) - y(i-2)) / dx^3;
    QM=((y(i-1)^3 + y(i)^3)/6)*yx3m + ((y(i-1)^2 + ...
    y(i)^2)/2)*alpha*xm*exp(-alpha*xm^2);
    ht0(i) = - (QP - QM) / dx;
end
ym = y(2);
ymm = y(3);
i = 1;
xp = L*(i-0.5)/N;
yx3p = (y(i+2) - 3*y(i+1) + 3*y(i) - ym) / dx^3;
QP=((y(i+1)^3 + y(i)^3)/6)*yx3p + ((y(i+1)^2 + ...
y(i)^2)/2)*alpha*xp*exp(-alpha*xp^2);
xm = L*(i-1.5)/N;
yx3m = (y(i+1) - 3*y(i) + 3*ym - ymm) / dx^3;
QM=((ym^3 + y(i)^3)/6)*yx3m + ((ym^2 + ...
y(i)^2)/2)*alpha*xm*exp(-alpha*xm^2);
ht0(i) = - (QP - QM) / dx;
i = 2;
xp = L*(i-0.5)/N;
yx3p = (y(i+2) - 3*y(i+1) + 3*y(i) - y(i-1)) / dx^3;
QP=((y(i+1)^3 + y(i)^3)/6)*yx3p + ((y(i+1)^2 + ...
y(i)^2)/2)*alpha*xp*exp(-alpha*xp^2);
xm = L*(i-1.5)/N;
yx3m = (y(i+1) - 3*y(i) + 3*y(i-1) - ym) / dx^3;
QM=((y(i-1)^3 + y(i)^3)/6)*yx3m + ((y(i-1)^2 + ...
y(i)^2)/2)*alpha*xm*exp(-alpha*xm^2);
ht0(i) = - (QP - QM) / dx;
ht = ht0';
```

B.2   Gravity-driven flow

B.2.1   Unsteady evolution equation: travelling wave

```
function tw
global N L b
b = 0.01;

% spatial mesh, initial condition

N = 1000;
L = 20;
x = (L/N)*(0:N-1);
y0 = [ ones(1,N/2) b.*ones(1,N/2) ];

% sparsity matrix for the Jacobian

e = ones(N,1);
S = spdiags([e e e e e], -2:2,N,N);
t = 0:200:600;
options = odeset('RelTol',1e-4,'AbsTol',1e-20, ...
'JPattern',S,'BDF','on');
[t,h] = ode15s(@f,t,y0,options);

e =(h(4,:) - h(3,:))*(h(4,:) - h(3,:))';
e = (e*(L/N))^0.5

plot(x,h(4,:),'Color','black','LineWidth',2)
xlabel('$\hat{x}$','Interpreter','Latex','Fontsize',20)
ylabel('$h_0$','Interpreter', ...
'Latex','Fontsize',20,'Rotation',0)
axis([0 10 0 1.8])

%-----------------------------------------------------------

function ht = f(t,y)
global N L b
dx = L/(N+1);
ucl = (1-b^3)/(3*(1-b));

y(N+1) = b;
y(N+2) = y(N);
for i = 3:N
    yx3p = (y(i+2) - 3*y(i+1) + 3*y(i) - y(i-1)) / dx^3;
    QP=((y(i+1)^3 + y(i)^3)/6)*yx3p + ((y(i+1)^3 + ...
    y(i)^3)/6) - ucl*((y(i+1) + y(i))/2);
    yx3m = (y(i+1) - 3*y(i) + 3*y(i-1) - y(i-2)) / dx^3;
```

```
        QM=((y(i-1)^3 + y(i)^3)/6)*yx3m + ((y(i-1)^3 + ...
        y(i)^3)/6) - ucl*((y(i-1) + y(i))/2);
        ht0(i) = - (QP - QM) / dx;
end
ym = 1;
ymm = y(1);
i = 1;
yx3p = (y(i+2) - 3*y(i+1) + 3*y(i) - ym) / dx^3;
QP=((y(i+1)^3 + y(i)^3)/6)*yx3p + ((y(i+1)^3 + ...
y(i)^3)/6) - ucl*((y(i+1) + y(i))/2);
yx3m = (y(i+1) - 3*y(i) + 3*ym - ymm) / dx^3;
QM=((ym^3 + y(i)^3)/6)*yx3m +((ym^3 + ...
y(i)^3)/6)- ucl*((ym + y(i))/2);
ht0(i) = - (QP - QM) / dx;
i = 2;
yx3p = (y(i+2) - 3*y(i+1) + 3*y(i) - y(i-1)) / dx^3;
QP=((y(i+1)^3 + y(i)^3)/6)*yx3p + ((y(i+1)^3 + ...
y(i)^3)/6) - ucl*((y(i+1) + y(i))/2);
yx3m = (y(i+1) - 3*y(i) + 3*y(i-1) - ym) / dx^3;
QM=((y(i-1)^3 + y(i)^3)/6)*yx3m +((y(i-1)^3 + ...
y(i)^3)/6) - ucl*((y(i-1) + y(i))/2);
ht0(i) = - (QP - QM) / dx;
ht = ht0';
```

B.2.2   Steady state equation

```
function bvppf
global L b

L = 16;
b = 0.2;
a=0.001;
options = bvpset('RelTol', 1e-4, 'NMax',10000);
solinit = bvpinit(linspace(0,L,5),@initpf,a);
sol = bvp4c(@odepf,@bcpf,solinit,options);
xint = linspace(0,L,400);
Sxint = deval(sol,xint);
figure(1);
plot(xint,Sxint(1,:));

%-----------------------------------------------------

function dydx = odepf(x,y,a)
global L b
```

```
dydx = [ y(2)
y(3)
((1-b^3)/(((y(1)^2)*(1-b))))-((b*(1+b))/(y(1)^3))-1];

%-----------------------------------------------------------

function res = bcpf(ya,yb,a)
global L b
q = 0.5.*(2 - b - b^2)^(1/3);
res = [ ya(1)-1-a; ...
ya(2)-a*q; ...
ya(3) + 2*a*q^2; yb(1)-b ];

%-----------------------------------------------------------

function yinit = initpf(x)
global L b
yinit = [ 1; 0; 0 ];
```

B.3   Landau-Levich problem

```
function ll
a = 0.001;
options = odeset('RelTol',1e-6,'AbsTol',1e-8,'Events', @events);
[t,h,te,he,ie] = ode45(@f,[0 Inf],[1+a a a],options);
b = 3^(2/3)*he(3)/sqrt(2)
plot(t,h(:,3),'k')
axis([0 20 0 0.8])
xlabel('$\hat{\xi}$','Interpreter', 'LaTeX')
ylabel('$\hat{H}''$','Interpreter', 'LaTeX')
end

% ----------------------------------------------------

function z = f(t,y)
z = [ y(2)
y(3)
(y(1)-1)/(y(1)^3) ];
end

% ----------------------------------------------------

function [v,ist,dir] = events(t,y)
v = (1/y(1)^2)-0.000001;
ist = 1;
dir = 0;
end
```

B.4 Electric double layers

```
eta = linspace(0,2,200);
eta0 = 0.5;
phi0 = asin(exp((eta-eta0)./2));
for i = 1:200
    s = exp((eta(i)-eta0)/2);
    h(i) = 2.*exp(-eta(i)/2)*(mfun('EllipticF',1, ...
    exp(-eta(i))) - mfun('EllipticF',s,exp(-eta(i))));
end
plot(h,eta,'k')

h0 = linspace(0,6,200);
etaDH = eta0./cosh(h0);
hold on
plot(h0,etaDH,'--k')
```

B.5 Fourier series solution for structured surfaces

```
delta = 0.5;
L = 2;
N=1000;
alh = zeros(N-1,1);
bh = zeros(N-1,1);
gh = zeros(N,1);

for n = 1:N-1
    kn = 2*pi*n/L;
    alh(n) = kn*tanh(kn)*(1-exp(-2*kn))/2;
    bh(n) = (kn*exp(-kn)/cosh(kn))-0.5*(1-exp(-2*kn));
end
r(1)= - pi*delta;
for i = 2:N
    r(i)= - sin(pi*(i-1)*delta)/(i-1);
end

% the matrix

Ah(1,1) = 2*pi*(1-delta);
for i = 2:N
    Ah(i,1) = - (2./(i-1))*sin(pi*delta*(i-1));
end
for i=1:N
```

```
    for j = 2:N
      if j ~= i
        Ah(i,j) = (alh(j-1)-bh(j-1))*...
        ((sin(pi*delta*(i+j-2))/(i+j-2))+...
        (sin(pi*delta*(i-j))/(i-j)));
      end
      if j == i
        Ah(i,i) = pi*(alh(i-1)*delta + ...
        bh(i-1)*(1-delta)) + ((alh(i-1)-bh(i-1))/...
        (2*(i-1)))*((alh(i-1)-bh(i-1))/(2*(i-1)))*...
        sin(2*pi*delta*(i-1));
      end
    end
end

gh = Ah\r';

% perturbation stream function

x = linspace(0,L);
y = linspace(0,1);
psi = zeros(100);
for i = 1:100
    for j = 1:100
      for n = 1:N-1
        kn = 2*pi*n/L;
        psi(i,j) = psi(i,j) + 0.5*gh(n+1)*...
        (exp(kn*(y(j)-1)) - exp(-kn*(y(j)+1))-...
        y(j)*tanh(kn)*(exp(kn*(y(j)-1)) ...
        + exp(-kn*(y(j)+1))))*cos(kn*x(i));
      end
    end
end

% contour plot

contour(x,y,psi',20,'k')
```

B.6   Local flux at the evaporating meniscus (from [103])

```
function mybvp
global Tw K delta ep haf

Tw = 1.4;
```

```
K = 2.0;
delta = 0.04;
ep = 10^(-3);
haf = ((delta*ep)/(Tw - 1))^(1/3);
xmax = 2.5;
pc=.01;
solinit = bvpinit(linspace(0,xmax,100),@myinit,pc);
sol = bvp4c(@myode,@mybc,solinit);
xint = linspace(0,xmax);
Sxint = deval(sol,xint);
figure(1);
xint=linspace(0,2.5);
plot(xint,[Sxint(1,:);Sxint(3,:)]);
axis([0 2.5 0 5.0]);
title('Interface shape');
xlabel('x');
ylabel('y');
h = Sxint(1,:);
hxx = Sxint(3,:);
Jint = (-delta.*hxx - delta.*ep.*(h.^(-3))) ...
+ Tw - 1)./(K + h);
figure(2); plot(xint,Jint); axis([-1 2.5 0 0.15]);
title('Evaporative flux');
xlabel('x');
ylabel('J');

%-----------------------------------------------------------

function dydx = myode(x,y,pc)
global Tw K delta ep haf
T = Tw - 1;
T1 = 1/T^(1/3);
dydx = [ y(2)
y(3)
y(4)
(3/(y(1)^3))*(((delta*y(3) + delta*ep*y(1)^(-3)...
- Tw + 1)/(K + y(1))) - y(1)^2*y(2)*y(4) - ...
((ep*y(2)^2)/y(1)^2) + (ep*y(3)/y(1)))];

%-----------------------------------------------------------

function res = mybc(ya,yb,pc)
global Tw K delta ep haf

ko = 1 + (pi/4)^(1/2);
mu = (3./((K+haf)*(haf^3)))^(1/2);
nu = (1./(haf^2))*(3.*ep)^(1/2);
```

```
res = [ ya(1) - haf - pc*haf - haf*.01;...
ya(2) - mu*pc*haf - nu*haf*.01; ...
ya(3) - mu^2*pc*haf - nu^2*haf*.01; ...
ya(4) - mu^3*pc*haf - nu^3*haf*.01; yb(3)- ko ];

%-------------------------------------------------------

function yinit = myinit(x)
global Tw K delta ep haf

yinit = [ x^2+haf; 2*x; 2; 0 ];
```

B.7   Surfactant spreading on initially uniform liquid film

```
function surf
global N L Pe
alpha = 100;
Pe = 1000;

% spatial mesh, initial condition

N = 1000;
NT = 2*N;
L = 40.0;
x = (L/N)*(0:N-1);
y0 = ones(1,NT);
for i = 1:N
    y0(N+i) = exp(-alpha.*(x(i)^2));
end

% time stepping

t = 0:200:600;
options = odeset('RelTol',1e-4,'AbsTol',1e-20,...
'BDF','on');
[t,yf] = ode15s(@f,t,y0,options);
yf2 = yf(2,:);
hf2 = yf2(:,1:N);
yf3 = yf(3,:);
hf3 = yf3(:,1:N);
yf4 = yf(4,:);
hf4 = yf4(:,1:N);
plot(x,hf2,'Color','black','LineWidth',2)
hold on
plot(x,hf3,'--','Color','black','LineWidth',2)
```

```
plot(x,hf4,'-.', 'Color','black','LineWidth',2)
xlabel('$x$','Interpreter','Latex','Fontsize',20)
ylabel('$h$','Interpreter','Latex','Fontsize',...
20,'Rotation',0)
legend('t=200','t=400','t=600')
axis([0 30 0.6 1.4])

%-----------------------------------------------------
function yt = f(t,y)
global N L Pe

dx = L/N;
for i = 1:N
    h(i) = y(i);
    G(i) = y(N+i);
end
h(N+1) = 1;
h(N+2) = h(N);
G(N+1) = 0;
G(N+2) = G(N);
for i = 3:N
    hx3p = (h(i+2) - 3*h(i+1) + 3*h(i) - h(i-1))/dx^3;
    Gxp = (G(i+1)-G(i))/dx;
    QP = -((h(i+1)^3 + h(i)^3)/6)*hx3p +...
    ((h(i+1)^2 + h(i)^2)/4)*Gxp;
    hx3m = (h(i+1) - 3*h(i) + 3*h(i-1) - h(i-2))/dx^3;
    Gxm = (G(i)-G(i-1))/dx;
    QM=-((h(i-1)^3 + h(i)^3)/6)*hx3m +...
    ((h(i-1)^2 + h(i)^2)/4)*Gxm;
    ht0(i) = (QP - QM) / dx;
    QGP = ((1./Pe)+((h(i+1)*G(i+1) + h(i)*G(i))/2)))...
    *GXP - ((h(i+1)^2*G(i+1) + h(i)^2*G(i))/4)*hx3p;
    QGM = ((1./Pe)+((h(i)*G(i) + h(i-1)*G(i-1))/2)))...
    *Gxm - ((h(i-1)^2*G(i-1) + h(i)^2*G(i))/4)*hx3m;
    ht0(N+i)= (QGP - QGM) / dx;
end
hm = h(2);
hmm = h(3);
Gm = G(2);
i = 1;
hx3p = (h(i+2) - 3*h(i+1) + 3*h(i) - hm) / dx^3;
Gxp = (G(i+1)-G(i))/dx;
QP=-((h(i+1)^3 + h(i)^3)/6)*hx3p + ...
((h(i+1)^2 + h(i)^2)/4)*Gxp;
xm=L*(i-1-0.5)/N;
```

```
hx3m = (h(i+1) - 3*h(i) + 3*hm - hmm) / dx^3;
Gxm = (G(i)-Gm)/dx;
QM=-((hm^3 + h(i)^3)/6)*hx3m +((hm^2 + ...
h(i)^2)/4)*Gxm;
ht0(i) = (QP - QM) / dx;
QGP = ((1./Pe)+((h(i+1)*G(i+1) + h(i)*G(i))/2)))...
*Gxp - ((h(i+1)^2*G(i+1) + h(i)^2*G(i))/4)*hx3p;
QGM = ((1./Pe)+((h(i)*G(i) + hm*Gm)/2))*Gxm - ...
((hm^2*Gm + h(i)^2*G(i))/4)*hx3m;
ht0(N+i)= (QGP - QGM) / dx;
i = 2;
hx3p = (h(i+2) - 3*h(i+1) + 3*h(i) - h(i-1))/dx^3;
Gxp = (G(i+1)-G(i))/dx;
QP = -((h(i+1)^3 + h(i)^3)/6)*hx3p + ...
((h(i+1)^2 + h(i)^2)/4)*Gxp;
hx3m = (h(i+1) - 3*h(i) + 3*h(i-1) - hm) / dx^3;
Gxm = (G(i)-G(i-1))/dx;
QM = -((h(i-1)^3 + h(i)^3)/6)*hx3m + ...
((h(i-1)^2 + h(i)^2)/4)*Gxm;
ht0(i) = (QP - QM) / dx;
QGP = ((1./Pe)+((h(i+1)*G(i+1) + h(i)*G(i))/2)))...
*Gxp - ((h(i+1)^2 + h(i)^2)/4)*hx3p;
QGM = ((1./Pe)+((h(i)*G(i) + h(i-1)*G(i-1))/2)))...
*Gxm - ((h(i-1)^2*G(i-1) + h(i)^2*G(i))/4)*hx3m;
ht0(N+i) = (QGP - QGM) / dx;
yt = ht0';
```

References

1. Acheson DJ (1990) Elementary fluid mechanics, Oxford University Press, Oxford
2. Adamson AW, Gast AP (1997) Physical chemistry of surfaces, 6th Ed., Wiley, New York
3. Ajaev VS, Brutin D, Tadrist L (2010) Evaporation of ultra-thin liquid films into air, Microgravity Sci. Tech. 22:441-446
4. Ajaev VS, Homsy GM (2001) Three-dimensional steady vapor bubbles in rectangular microchannels, J. Colloid Interface Sci., 244:180-189
5. Ajaev VS, Homsy GM (2006) Modeling shapes and dynamics of confined bubbles, Annu. Rev. Fluid Mech., 38:277-307
6. Ajaev VS, Homsy GM, Morris SJS (2002) Dynamic response of geometrically constrained vapor bubbles, J. Colloid Interface Sci. 254:346-354
7. Ajaev VS, Willis DA (2003) Thermocapillary flow and rupture in films of molten metal on a substrate, Phys. Fluids, 15:3144-3150
8. Andelman, D (2006) Introduction to electrostatics in soft and biological matter, In: Proceedings of the Nato ASI & SUSSP on soft condensed matter physics in molecular and cell biology, W. Poon and D. Andelman (Eds.), Taylor & Francis, New York
9. Anderson DM, Davis SH (1995) The spreading of volatile liquid droplets on heated surfaces, Phys. Fluids 7:248-265.
10. Bahga SS, Vinogradova OI, Bazant MZ (2010) Anisotropic electroosmotic flow near super-hydrophobic surfaces, J. Fluid Mech. 644:245-255
11. Barash LYu, Bigioni TP, Vinokur VM, Shchur LN (2009) Evaporation and fluid dynamics of a sessile drop of capillary size, Phys. Rev. E 79:046301.1-16
12. Batchelor GK (1967) An introduction to fluid dynamics, Cambridge University Press, Cambridge
13. Bazant MZ, Vinogradova OI (2008) Tensorial hydrodynamic slip, J. Fluid Mech. 613:125-134
14. Bender C, Orszag S (1999) Advanced mathematical methods for scientists and engineers – I, Springer, New York
15. Belyaev AV, Vinogradova OI (2010) Effective slip in pressure-driven flow past super-hydrophobic stripes, J. Fluid Mech. 652:489-499
16. Berthier J (2007) Microdrops and digital microfluidics, William Andrew, Norwich, New York
17. Bertozzi AL, Brenner MP (1997) Linear stability and transient growth in driven contact lines, Phys. Fluids 9:530-539
18. Bico J, Thiele U, Quéré D (2002) Wetting of textured surfaces, Coll. Surf. A 206:41-46
19. Blossey R (2003) Self-cleaning surfaces – virtual realities, Nature Materials 2:301-306
20. Bonn D, Eggers J, Indekeu J, Meunier J, Rolley E (2009) Wetting and spreading, Rev. Mod. Phys. 81:739-804
21. Brennen C (1995) Cavitation and bubble dynamics, Oxford Univ. Press, Oxford

22. Brenner MP, Hilgenfeld S, Lohse D (2002) Single-bubble sonoluminescence, Rev. Mod. Phys. 74:425-484

23. Bretherton FP (1961) The motion of long bubbles in tubes, J. Fluid Mech. 10:166-188

24. Brown PN, Byrne CD, Hindmarsh AC (1989) VODE: A Variable Coefficient ODE Solver, SIAM J. Sci. Stat. Comput. 10:1038–1051

25. Bruus H (2008) Theoretical microfluidics, Oxford University Press, Oxford

26. Buck AL (1981) New equations for computing vapor pressure and enhancement factor, J. Appl. Meteor. 20:1527-1532

27. Burelbach JP, Bankoff SG, Davis SH (1988) Nonlinear stability of evaporating/condensing liquid films, J. Fluid Mech. 195:463

28. Butt H-J, Kappl M (2010) Surface and interfacial forces, Wiley-VCH, Weinheim

29. Canuto C, Hussaini MY, Quarteroni A, Zang TA (2006) Spectral methods: fundamentals in single domains, Springer, Berlin

30. Carey VP (2007) Liquid-vapor phase change phenomena Taylor & Francis, Oxford

31. Cassie ABD, Baxter S (1944) Wettability of porous surfaces, Trans. Faraday Soc 40:546-551

32. Cermelli P, Fried E, Gurtin M (2005) Transport relations for surface integrals arising in the formulation of balance laws for evolving interfaces, J. Fluid Mech. 544:339-351

33. Chang H-C, Yeo LY (2010) Electrokinetically driven microfluidics and nanofluidics, Cambridge University Press, Cambridge

34. Coffey T, Kelley CT, Keyes DE (2003) Pseudotransient continuation and differential-algebraic equations, SIAM J. Sci. Comput. 25:553-569

35. Colinet P, Haut B (2005) Surface-tension-driven instabilities of a pure liquid layer evaporating into an inert gas, J. Coll. Interface Sci. 285:296-305

36. Concus P, Finn R (1969) On the behavior of a capillary surface in a wedge, Proc. Natl. Acad. Sci. 63:292-299.

37. Connor JN, Horn RG (2001) Measurement of aqueous film thickness between charged mercury and mica surfaces: a direct experimental probe of the Poisson-Boltzmann distribution, Langmuir 17:7194-7197

38. Cottin-Bizonne C, Cross B, Steinberger A, Charlaix E (2005) Boundary slip on smooth hydrophobic surfaces: intrinsic effects and possible artifacts, Phys. Rev. Lett. 94:056102

39. Cox RG (1986) The dynamics of the spreading of liquids on a solid surface. Part I. Viscous flow, J. Fluid Mech. 168:169-194

40. Craster RV, Matar OK (2009) Dynamics and stability of thin liquid films. Rev. Mod. Phys. 81:1131-1198

41. DasGupta S, Schonberg JA, Wayner PC (1993) Investigation of an evaporating extended meniscus based on the augmented Young-Laplace equation, ASME J. Heat Transfer 115:201-208

42. Davis SH (1987) Thermocapillary instabilities, Annu. Rev. Fluid Mech. 19:403.

43. Deegan, RD, Bakajin O, Dupont TF, Huber G, Nagel SR, Witten, TA (2000) Contact line deposits in an evaporating drop, Phys. Rev. E 62:756-765

44. Derjaguin BV, Kussakov MM (1939) Anomalous properties of thin polymolecular films, Acta Physicochim. URSS 10:153-174

45. Derjaguin BV, Churaev NV, Muller VM (1987) Surface forces. Plenum, New York

46. Dunn G, Wilson SK, Duffy BR, David S and Sefiane K (2009) The strong influence of substrate conductivity on droplet evaporation, J. Fluid Mech. 623:329-351

47. Dzyaloshinskii IE, Lifshitz EM, Pitaevskii LP (1960) Van der Waals forces in liquid films, JETP 10:161-170

48. Edmonstone BD, Craster RV, Matar OK (2006) Surfactant-induced fingering phenomena beyond the critical micelle concentration. J. Fluid Mech. 564:105-138

49. Eggers J. Evans R. (2004) Comment on "Dynamic wetting by liquids of different viscosity", by T.D. Blake and Y.D. Shikhmurzaev. Journal of Colloid and Interface Science, 280(2), 537-538

50. Eggers J, Stone HA (2004) Characteristic lengths at moving contact lines for a perfectly wetting fluid: the influence of speed on the dynamic contact angle, J. Fluid Mech. 505:309-321

51. Eres MH, Schwartz LW, Roy LV (2000) Fingering phenomena for driven coating films, Phys. Fluids 12:1278-1295
52. Feuillebois F, Bazant MZ, Vinogradova OI (2009) Effective slip over superhydrophobic surfaces in thin channels, Phys. Rev. Lett. 102:026001
53. Finn R (1986) Equilibrium capillary surfaces, Springer, New York
54. Gao X, Yan X, Yao X, Xu L, Zhang K, Zhang J, Yang B, and Jiang L (2007) The dry-style antifogging properties of mosquito compound eyes and artificial analogues prepared by soft lithography, Adv. Materials 19:2213-2217
55. de Gennes PG (1985) Wetting: statics and dynamics, Rev. Mod. Phys. 57: 827-863
56. de Gennes PG, Brochard-Wyart F, Quere D. (2004) Capillarity and wetting phenomena: drops, bubbles, pearls, waves, Springer, New York
57. Ghosal S (2006) Electrokinetic flow and dispersion in capillary electrophoresis, Annu. Rev. Fluid Mech. 38:309-338
58. Goodwin R, Homsy GM (1991) Viscous flow down a slope in the vicinity of a contact line, Phys. Fluids A 515-528
59. Greenspan HP (1978) On the motion of a small viscous droplet that wets a surface, J. Fluid Mech. 84:125-143
60. Grigoriev RO (2005) Transient growth in driven contact lines, Physica D 209:105-116
61. Guenther A, Jensen KF (2006) Multiphase microfluidics: from flow characteristics to chemical and materials synthesis, Lab-on-a-Chip, 6:1487-1503
62. Hunter RJ (2001) Foundations of colloid science, Oxford Univ. Press, Oxford
63. Hayes RA, Feenstra BJ (2003) Video-speed electronic paper based on electrowetting, Nature 425:383-385
64. Hardt S, Shoenfield W, Editors (2009) Microfluidic technologies for miniaturized analysis systems, Springer New York
65. Hewitt D, Fornasiero D, Ralston J, Fisher LR (1993) Aqueous film drainage at the quartz/water/air interface, J. Chem Soc. Faraday Trans. 89:817-822
66. Hoffman RL (1979) A study of the advancing interface. I. Interface shape in liquid-gas systems, J. Colloid Interface Sci. 50:228-241
67. Homsy GM (1987) Viscous fingering in porous media, Annu. Rev. Fluid Mech. 19:271-311
68. Huh C, Scriven LE (1971) Hydrodynamic model of steady movement of a solid/liquid/fluid contact line, J. Colloid Interface Sci. 35:85-101
69. Huppert H (1982) Flow and instability of a viscous current down a slope, Nature, 300:427-429
70. Dussan V, EB (1979) On the spreading of liquids on solid surfaces: static and dynamic contact lines, Annu. Rev. Fluid Mech. 11:371-400
71. Holmes MH (1995) Introduction to perturbation methods, Springer, New York
72. Israelachvili JN (2011) Intermolecular and surface forces, 3rd Edition, Elsevier, Amsterdam
73. Jensen JB, Grotberg JB (1992) Insoluble surfactant spreading on a thin viscous film: shock evolution and film rupture, J. Fluid Mech. 240: 259-288
74. Jensen JB, Grotberg JB (1993) The spreading of heat or soluble surfactant along a thin liquid film, Phys. Fluids A 5:58-68
75. Kelley CT (1995) Iterative methods for linear and nonlinear equations, SIAM, Philadelphia
76. Kim IY, Wayner PC, Jr (1996) Shape of an evaporating completely wetting extended meniscus, J. Thermophys. Heat Transfer, 10:320-325
77. Kirby BJ (2010) Micro- and nanoscale fluid mechanics: transport in microfluidic systems, Cambridge University Press, Cambridge
78. Kogan MN (1969) Rarefied gas dynamics, Plenum Press, New York
79. Kondepudi D, Prigogine I (1998) Modern Thermodynamics, Wiley, New York
80. Kondic L, Diez J (2001) Pattern formation in the flow of thin films down an incline: Constant flux configuration, Phys. Fluids 13:3168-3184
81. Labuntsov DA, Kryukov AP (1979) Analysis of intensive evaporation and condensation, Int. J. Heat Mass Transfer, 22:989-1002
82. Lafuma A, Quéré D (2003) Superhydrophobic states, Nature Mater. 2:457-460

83. Landau L, Levich B (1942) Dragging of a liquid by a moving plate, Acta Physicochimica (USSR), 17:42-54
84. Lifshitz EM, Pitaevskii LP (1981) Physical kinetics, Butterworth-Heinemann, Oxford
85. Lauga E, Brenner MP, Stone HA (2007) Microfluidics: The no-slip boundary condition, *in* Handbook of Experimental Fluid Dynamics (Chapter 19), C. Tropea, A. Yarin, J. F. Foss (Eds.), Springer
86. Lauga E, Stone HA (2003) Effective slip in pressure-driven Stokes flow, J. Fluid Mech., 489:55-77
87. Leal LG (2007) Advanced transport phenomena: fluid mechanics and convective transport processes, Cambridge Univ. Press, Cambridge
88. Levich VG (1962) Physicochemical hydrodynamics, Prentice Hall, Englewood Cliffs
89. Li D (2004) Electrokinetics in microfluidics, Elsevier, Amsterdam
90. Li H, Yoda M (2010) An experimental study of slip considering the effects of non-uniform colloidal tracer distributions, J. Fluid Mech., 662:269-287
91. Manica R, Chan DYC (2011) Drainage of the air-water-quartz film: theory and experiment, Physical Chem. Chem. Physics 13:1434-1439
92. Moosman S, Homsy GM (1980) Evaporating menisci of wetting fluids, J. Colloid Interface Sci., 73:212-223
93. Morris SJS (2001) Contact angles for evaporating liquids predicted and compared with existing experiments, J. Fluid Mech. 432:1-30
94. Moyle DT, Chen M-S, Homsy GM (1999) Nonlinear rivulet dynamics during unstable wetting flows, Int. J. Multiphase Flow 25:1243-1262
95. Mugele F, Baret J.-C. (2005) Electrowetting: from basics to applications, J. Phys.:Condens. Matter 17: R705-R774
96. O'Neill B (2006) Elementary differential geometry, 2nd Edition, Academic Press (Elsevier) Burlington, MA
97. Oron A, Davis SH, Bankoff SG (1997) Long-scale evolution of thin liquid films. Rev. Mod. Phys. 69:931-980.
98. Parker AR, Lawrence CR (2001) Water capture by a desert beetle, Nature 414:33-34
99. Petzold LR (1982) A description of DASSL: a differential/ algebraic system solver, SAND82-8637, Sandia National Laboratories, NM
100. Philip R (1972) Flows satisfying mixed no-slip and no-shear conditions, Z. Angew Math Phys 23:353-372
101. Potash M, Wayner PC (1972) Evaporation from a two-dimensional extended meniscus, Int. J. Heat and Mass Transfer 15: 1851-1863
102. Probstein RF (1989) Physicochemical hydrodynamics: An introduction, Butterworth-Heinemann, Boston
103. Quach HR, Ajaev VS (2005) Numerical computation of local vapor-liquid interface shape and heat transfer near steady contact line on heated surface, SMU Applied Mathematics Technical Report 2005-04, Dallas
104. Quéré, D (2005) Non-sticking droplets, Rep. Prog. Phys. 68:2495-2532
105. Quéré D (2008) Wetting and roughness, Annu. Rev. Mater. Res. 38:71-99
106. Rayleigh, Lord (1917) On the pressure developed in a liquid during the collapse of a spherical cavity. Phil. Mag. 34:94-98.
107. Richard D, Quéré, D (2000) Bouncing water drops, Europhys. Lett. 50:769-775
108. Rose JW (2000) Accurate approximate equations for intensive sub-sonic evaporation, Int. J. Heat Mass Transfer 43:3869-3875
109. Sbragaglia M, Prosperetti A (2007) A note on the effective slip properties for microchannel flows with ultrahydrophobic surfaces, Phys. Fluids 19:043603
110. Shampine LF, Gladwell I, Thompson S (2003) Solving ODEs with MATLAB, Cambridge Univ. Press, Cambridge
111. Shikhmurzaev YD (2008) Capillary flows with forming interfaces, Chapman & Hall/CRC, Boca Raton

112. Shikhmurzaev YD (2004) Response to the comment on [J. Colloid Interface Sci. 253 (2002) 196] by J. Eggers and R. Evans, J. Colloid. and Interface Sci. 280 539-541
113. Schrage RW (1953) A theoretical study of interphase mass transfer, Columbia University Press, New York
114. Silvi N, Dussan V EB (1985) On the rewetting of an inclined solid surface by a liquid, Phys. Fluids 28:5-7
115. Slattery JC, Sagis LMC, Oh E-S (2006) Interfacial transport phenomena, Springer, New York
116. Sneddon IN (1966) Mixed boundary value problems in potential theory, North Holland, Amsterdam
117. Sodtke C, Ajaev VS, Stephan P (2008) Dynamics of volatile liquid droplets on heated surfaces: theory versus experiment, J. Fluid Mech. 610:343-362
118. Sone Y (2007) Molecular gas dynamics: theory, techniques, and applications, Birkhauser, Boston
119. Spaid MA, Homsy GM (1996) Stability of Newtonian and viscoelastic dynamic contact lines, Phys. Fluids 8:460-478
120. Squires TM, Quake SR (2005) Microfluidics: Fluid physics at the nanoliter scale, Rev. Mod. Phys. 77:977-1026
121. Squires TM (2008) Electrokinetic flows over inhomogeneously slipping surfaces, Phys. Fluids 20 092105
122. Starov VM, Velarde MG, Radke CJ (2007) Wetting and spreading dynamics, CRC Press, Boca Raton
123. Stone HA, Strook AD, Ajdari A (2004) Engineering flows in small devices: Microfluidics toward a lab-on-a-chip, Annu. Rev. Fluid Mech. 36:381-411
124. Stone HA (1990) A simple derivation of time-dependent convective-diffusion equation for surfactant transport along a deforming interface, Phys. Fluids A 2:111-112
125. Subramanian RS, Balasubramaniam R (2001) The motion of bubbles and drops in reduced gravity, Cambridge University Press, Cambridge
126. Sultan E, Boudaoud A, Ben Amar M (2005) Evaporation of a thin film: Diffusion of the vapour and Marangoni instabilities, J. Fluid Mech. 543:183-202
127. Tabeling P (2005) Introduction to microfluidics, Oxford University Press, Oxford
128. Tanner LH (1979) The spreading of silicone oil drops on horizontal surfaces, J. Phys. D: Appl. Phys. 12:1473-1484
129. Teo CJ, Khoo BC (2009) Analysis of Stokes flow in microchannels with superhydrophobic surfaces containing a periodic array of micro-grooves, Microfluid Nanofluid 7:353-382
130. Thomas JW (1995) Numerical partial differential equations: finite difference methods, Springer, New York
131. Tuck EO, Schwartz LW (1990) A numerical and asymptotic study of some third-order ordinary differential equations relevant to draining and coating flows, SIAM Review, 32:453-469
132. Vinogradova OI (1999) Slippage of water over hydrophobic surfaces, Int. J. Mineral Processing, 56:31-60
133. Vinogradova OI, Belyaev AV (2011) Wetting, roughness and flow boundary conditions, J. Phys.: Condens. Matter 23:184104
134. Voinov OI (1976) Hydrodynamics of wetting, Izv. Akad Nauk SSSR Mekh. Zhid. i Gaza 5:76-84
135. Ward C, Fang C (1999) Expression for predicting liquid evaporation flux: the statistical rate theory approach, Phys. Rev. E 59:429-440
136. Wayner PC (1999) Intermolecular forces in phase-change heat transfer: 1998 Kern award review, AIChE J 45:2055-2068
137. Wenzel RN (1936) Resistance of solid surfaces to wetting by water, Ind. Eng. Chem. 28:988-994
138. Williams MB, Davis SH (1982) Nonlinear theory of film rupture, J. Colloid Interface Sci. 90:220-228

139. Willis DA, Xu X (2000) Transport phenomena and droplet formation during pulsed laser interaction with thin films, ASME J. Heat Tranfer, 122:763-770

140. Witelski TP, Bowen M (2003) ADI schemes for higher-order nonlinear diffusion equations, Appl. Num. Math. 45:331-351

141. Wong H, Rumschitzki D, Maldarelli C (1996) On the surfactant mass balance at a deforming fluid interface, Phys. Fluids, 8:3203-3204

142. Wong H, Morris S, Radke CJ (1992) Three-dimensional menisci in polygonal capillaries, J. Colloid Interface Sci., 148:317-336

143. Wong H, Morris S, Radke CJ (1995) The motion of long bubbles in polygonal capillaries. Part I. Thin films. J. Fluid Mech., 292:71-94

144. Wong H, Morris S, Radke CJ (1995) The motion of long bubbles in polygonal capillaries, Part II. Drag, fluid pressure, and fluid flow. J. Fluid Mech. 292:95-110

145. Yang L, Homsy GM (2008) Experimental study of vapor bubbles in small-sized channels, J. Colloid Interface Sci. 317:235-240

146. Ytrehus T (1977) Theory and experiments on gas kinetics in evaporation, In: Rarefied gas dynamics, edited by J. L. Potter, AIAA, New York, 1197-1211

147. Zhang WW, Lister JR (1999) Similarity solutions for van der Waals rupture of a thin film on a solid substrate, Phys. Fluids 11:2454-2462

148. Zhornitskaya L, Bertozzi AL (2000) Positivity-preserving schemes for lubrication-type equations, SIAM J. Numer. Anal. 37:523–555

Index

V.S. Ajaev, *Interfacial Fluid Mechanics: A Mathematical Modeling Approach*,
DOI 10.1007/978-1-4614-1341-7, © Springer Science+Business Media, LLC 2012